C++
for
Financial
Mathematics

CHAPMAN & HALL/CRC
Financial Mathematics Series

Aims and scope:
The field of financial mathematics forms an ever-expanding slice of the financial sector. This series aims to capture new developments and summarize what is known over the whole spectrum of this field. It will include a broad range of textbooks, reference works and handbooks that are meant to appeal to both academics and practitioners. The inclusion of numerical code and concrete real-world examples is highly encouraged.

Series Editors

M.A.H. Dempster
*Centre for Financial Research
Department of Pure
Mathematics and Statistics
University of Cambridge*

Dilip B. Madan
*Robert H. Smith School
of Business
University of Maryland*

Rama Cont
*Department of Mathematics
Imperial College*

Published Titles

American-Style Derivatives; Valuation and Computation, *Jerome Detemple*

Analysis, Geometry, and Modeling in Finance: Advanced Methods in Option
  Pricing, *Pierre Henry-Labordère*

C++ for Financial Mathematics, *John Armstrong*

Commodities, *M. A. H. Dempster and Ke Tang*

Computational Methods in Finance, *Ali Hirsa*

Counterparty Risk and Funding: A Tale of Two Puzzles, *Stéphane Crépey and
  Tomasz R. Bielecki, With an Introductory Dialogue by Damiano Brigo*

Credit Risk: Models, Derivatives, and Management, *Niklas Wagner*

Engineering BGM, *Alan Brace*

Financial Mathematics: A Comprehensive Treatment, *Giuseppe Campolieti and
  Roman N. Makarov*

The Financial Mathematics of Market Liquidity: From Optimal Execution to
  Market Making, *Olivier Guéant*

Financial Modelling with Jump Processes, *Rama Cont and Peter Tankov*

Interest Rate Modeling: Theory and Practice, *Lixin Wu*

Introduction to Credit Risk Modeling, Second Edition, *Christian Bluhm,
  Ludger Overbeck, and Christoph Wagner*

An Introduction to Exotic Option Pricing, *Peter Buchen*

Introduction to Risk Parity and Budgeting, *Thierry Roncalli*

Introduction to Stochastic Calculus Applied to Finance, Second Edition,
  Damien Lamberton and Bernard Lapeyre

Monte Carlo Methods and Models in Finance and Insurance, *Ralf Korn, Elke Korn,
  and Gerald Kroisandt*

Monte Carlo Simulation with Applications to Finance, *Hui Wang*

Nonlinear Option Pricing, *Julien Guyon and Pierre Henry-Labordère*
Numerical Methods for Finance, *John A. D. Appleby, David C. Edelman, and John J. H. Miller*
Option Valuation: A First Course in Financial Mathematics, *Hugo D. Junghenn*
Portfolio Optimization and Performance Analysis, *Jean-Luc Prigent*
Quantitative Finance: An Object-Oriented Approach in C++, *Erik Schlögl*
Quantitative Fund Management, *M. A. H. Dempster, Georg Pflug, and Gautam Mitra*
Risk Analysis in Finance and Insurance, Second Edition, *Alexander Melnikov*
Robust Libor Modelling and Pricing of Derivative Products, *John Schoenmakers*
Stochastic Finance: An Introduction with Market Examples, *Nicolas Privault*
Stochastic Finance: A Numeraire Approach, *Jan Vecer*
Stochastic Financial Models, *Douglas Kennedy*
Stochastic Processes with Applications to Finance, Second Edition, *Masaaki Kijima*
Stochastic Volatility Modeling, *Lorenzo Bergomi*
Structured Credit Portfolio Analysis, Baskets & CDOs, *Christian Bluhm and Ludger Overbeck*
Understanding Risk: The Theory and Practice of Financial Risk Management, *David Murphy*
Unravelling the Credit Crunch, *David Murphy*

Proposals for the series should be submitted to one of the series editors above or directly to:
CRC Press, Taylor & Francis Group
3 Park Square, Milton Park
Abingdon, Oxfordshire OX14 4RN
UK

Chapman & Hall/CRC FINANCIAL MATHEMATICS SERIES

C++ for Financial Mathematics

John Armstrong
King's College London, Strand, UK

CRC Press is an imprint of the
Taylor & Francis Group, an **informa** business

A CHAPMAN & HALL BOOK

CRC Press
Taylor & Francis Group
6000 Broken Sound Parkway NW, Suite 300
Boca Raton, FL 33487-2742

© 2017 by Taylor & Francis Group, LLC
CRC Press is an imprint of Taylor & Francis Group, an Informa business

No claim to original U.S. Government works

Printed on acid-free paper
Version Date: 20161202

International Standard Book Number-13: 978-1-4987-5005-9 (Hardback)

This book contains information obtained from authentic and highly regarded sources. Reasonable efforts have been made to publish reliable data and information, but the author and publisher cannot assume responsibility for the validity of all materials or the consequences of their use. The authors and publishers have attempted to trace the copyright holders of all material reproduced in this publication and apologize to copyright holders if permission to publish in this form has not been obtained. If any copyright material has not been acknowledged please write and let us know so we may rectify in any future reprint.

Except as permitted under U.S. Copyright Law, no part of this book may be reprinted, reproduced, transmitted, or utilized in any form by any electronic, mechanical, or other means, now known or hereafter invented, including photocopying, microfilming, and recording, or in any information storage or retrieval system, without written permission from the publishers.

For permission to photocopy or use material electronically from this work, please access www.copyright.com (http://www.copyright.com/) or contact the Copyright Clearance Center, Inc. (CCC), 222 Rosewood Drive, Danvers, MA 01923, 978-750-8400. CCC is a not-for-profit organization that provides licenses and registration for a variety of users. For organizations that have been granted a photocopy license by the CCC, a separate system of payment has been arranged.

Trademark Notice: Product or corporate names may be trademarks or registered trademarks, and are used only for identification and explanation without intent to infringe.

Visit the Taylor & Francis Web site at
http://www.taylorandfrancis.com

and the CRC Press Web site at
http://www.crcpress.com

Printed and bound in the United States of America by
Edwards Brothers Malloy on sustainably sourced paper

Contents

Introduction		**xvii**
1	**Getting Started**	**1**
	1.1 Installing your development environment	1
	1.1.1 For Windows	1
	1.1.2 For Unix	1
	1.1.3 For MacOS X	1
	1.2 Running an example program	2
	1.3 Compiling and running the code	3
	1.3.1 Compiling on Windows	4
	1.3.2 Compiling on Unix	6
	1.4 Understanding the example code	8
	1.5 Configuring the compiler	12
	1.6 Making decisions	13
	1.7 Exercises	14
	1.8 Summary	15
2	**Basic Data Types and Operators**	**17**
	2.1 Memory terminology	17
	2.2 Basic data types	18
	2.2.1 Integers	18
	2.2.2 Floating point numbers	20
	2.2.3 Booleans	20
	2.2.4 Characters	20
	2.3 Casting	22
	2.4 Memory addresses	26
	2.5 Operators	28
	2.5.1 The `sizeof` operator	28
	2.5.2 Mathematical operations	28
	2.5.3 Comparison operators	29
	2.5.4 Logical operators	29
	2.5.5 Bitwise operators	29
	2.5.6 Combining operators	30
	2.5.7 Assignment operators	30

vii

viii   *Contents*

| | | 2.5.8 | If statements revisited | 32 |
| | 2.6 | Summary | | 35 |

3 Functions   37

	3.1	The C++ function syntax	37
	3.2	Recursion	41
	3.3	Libraries	42
	3.4	Declaring and defining functions	42
	3.5	Functions that don't return a value	44
	3.6	Specifying default values	45
	3.7	Overloading functions	46
	3.8	Global and local variables	47
	3.9	Namespaces	48
	3.10	Summary	52

4 Flow of Control   55

	4.1	`while` loops	55
	4.2	`do-while` loops	57
	4.3	`for` loops	58
	4.4	`break`, `continue`, `return`	60
	4.5	`throw` statements	61
	4.6	`switch` statements	63
	4.7	Scope	65
	4.8	Flow of control in operators	65
		4.8.1 Short circuit evaluation	66
		4.8.2 The ternary operator	66
		4.8.3 The comma operator	67
	4.9	Summary	69

5 Working with Multiple Files   71

	5.1	The project FMLib	71
	5.2	Header files	72
	5.3	Creating our project	73
		5.3.1 Creating the first header file	73
		5.3.2 Some code that uses the functions	75
		5.3.3 Write the definitions	76
	5.4	How header files work	77
		5.4.1 The meaning of include	77
		5.4.2 Pragma once	77
		5.4.3 Information hiding	78
		5.4.4 Inline	80
	5.5	A complete example	81

Contents   ix

5.6	Summary	82

6   Unit Testing   85

6.1	A testing framework for C++	86
6.2	Macros	86
6.3	The macros in `testing.h`	87
	6.3.1 The `ASSERT` macro	87
	6.3.2 The `ASSERT_APPROX_EQUAL` macro	87
	6.3.3 The `INFO` macro	88
	6.3.4 The `DEBUG_PRINT` macro	88
	6.3.5 The TEST macro	89
6.4	Using `testing.h`	89
6.5	What have we gained?	91
6.6	Testing `normcdf`	92
6.7	Summary	94

7   Using C++ Classes   97

7.1	Vectors	97
7.2	Pass by reference and const	100
	7.2.1 Pass by reference	101
	7.2.2 The const keyword	102
	7.2.3 Pass by reference without `const`	104
7.3	Using `ofstream`	104
7.4	Working with `string`	106
7.5	Building strings efficiently	107
7.6	Writing a pie chart	108
	7.6.1 A web-based chart	109
	7.6.2 Create a header file	111
	7.6.3 Write a source file	112
	7.6.4 Enable testing in your files	112
	7.6.5 Write functions to generate the boiler plate	112
	7.6.6 Write a simple version of the chart data	113
	7.6.7 Write a test of what we've done so far	114
	7.6.8 Write the interesting code	114
	7.6.9 Testing the interesting code	115
	7.6.10 Wrap it all up into a single function	116
7.7	The architecture of the World Wide Web	117
7.8	Summary	121

8   User-Defined Types   123

8.1	Terminology	123
8.2	Writing your own class	124

Contents

8.2.1	Writing the declaration	124
8.2.2	Using a class	126
8.2.3	Passing objects between functions	127
8.2.4	How have classes helped?	127
8.3	Adding functions to classes	128
8.3.1	Using const on member functions	130
8.4	A financial example	131
8.4.1	What have we gained?	133
8.5	Recommendations on writing classes	134
8.6	Encapsulation	135
8.6.1	Implementing PieChart	137
8.6.2	Using PieChart	137
8.7	Constructors	138
8.7.1	Writing a default constructor	139
8.7.2	An alternative, and superior syntax	140
8.8	Constructors with parameters	141
8.9	Summary	144

9 Monte Carlo Pricing in C++   145

9.1	A function to simulate stock prices	146
9.2	Writing a Monte Carlo pricer	151
9.3	Generating random numbers for Monte Carlo	154
9.4	Summary	158

10 Interfaces   159

10.1	An interface for pricing options	159
10.2	Describing an interface in C++	161
10.3	Examples of interfaces	164
10.4	Interfaces in object-oriented programming	166
10.5	What's wrong with if statements?	168
10.6	An interface for integration	169
10.7	Summary	173

11 Arrays, Strings, and Pointers   175

11.1	Arrays, the C alternative to `vector`	176
11.2	Pointers	179
11.2.1	`new` and `delete`	179
11.2.2	Pointer operators	180
11.2.3	Looping with pointers	182
11.2.4	Using pointers in practice	185
11.3	Pointers to text	185
11.4	Pass by pointer	187

Contents

11.5	Don't return pointers to local variables	189
11.6	Using pointers to share data	191
	11.6.1 Sharing with shared_ptr	194
11.7	Sharing data with references	197
11.8	The C++ memory model	199
	11.8.1 The stack	200
	11.8.2 The heap	202
11.9	Summary	204

12 More Sophisticated Classes — 205

12.1	Inlining member functions	205
12.2	The `this` keyword	206
12.3	Inheritance	207
	12.3.1 What have we gained?	209
	12.3.2 Terminology	209
12.4	Overriding methods — the `virtual` keyword	210
	12.4.1 A note on the keyword `virtual`	211
12.5	Abstract functions =0	212
12.6	Multiple layers	212
	12.6.1 UML	213
	12.6.2 Another object hierarchy	215
	12.6.3 Multiple inheritance	215
	12.6.4 Calling superclass methods	216
12.7	Forward declarations and the structure of cpp files	217
12.8	The `static` keyword	218
12.9	The `protected` keyword	220
12.10	Summary	222

13 The Portfolio Class — 223

13.1	The `Priceable` interface	223
13.2	The `Portfolio` interface and implementation	224
	13.2.1 Implementation of `PortfolioImpl`	227
13.3	Testing	228
13.4	UML	230
13.5	Limitations	231
13.6	Summary	232

14 Delta Hedging — 233

14.1	Discrete-time delta hedging	233
14.2	Implementing the delta hedging strategy in C++	235
	14.2.1 Class declaration	235
	14.2.2 Implementation of `runSimulation`	237

xii   *Contents*

	14.2.3 Implementing the other methods of HedgingSimulator	238
	14.2.4 Changes to `CallOption`	240
14.3	Testing the simulation	241
14.4	Interpreting and extending our simulation	241
14.5	Summary	244

15 Debugging and Development Tools   245

15.1	Debugging strategies	245
	15.1.1 Unit tests	245
	15.1.2 Reading your code	246
	15.1.3 Logging statements	246
	15.1.4 Using a debugger	247
	15.1.5 Divide and conquer	247
15.2	Debugging with Visual Studio	248
	15.2.1 Obtaining a stack trace in Visual Studio	248
	15.2.2 Breakpoints and single stepping in Visual Studio . .	250
15.3	Debugging with GDB	252
	15.3.1 Using GDB to obtain a stack trace	253
	15.3.2 Breakpoints and single stepping with GDB	256
	15.3.3 Other commands and features	257
15.4	Other development tools and practices	258
	15.4.1 Version control	258
	15.4.2 Bug tracking	259
	15.4.3 Testing framework	259
	15.4.4 Automated build	260
	15.4.5 Continuous integration	261
	15.4.6 Logging	261
	15.4.7 Static analysis	261
	15.4.8 Memory-leak detection	262
	15.4.9 Profiling tools	262
	15.4.10 Example	263
15.5	Summary	264

16 A Matrix Class   267

16.1	Basic functionality of `Matrix`	267
16.2	The constructor and destructor of `Matrix`	269
	16.2.1 Virtual destructors	271
	16.2.2 When is a destructor needed?	272
	16.2.3 Additional constructors	273
16.3	Const pointers	274
16.4	Operator overloading	275
	16.4.1 Overloading + ...	275
	16.4.2 Overloading other arithmetic operators	277

Contents

	16.4.3 Overloading comparison operators	278
	16.4.4 Overloading the << operator	279
	16.4.4.1 Remarks on return by reference	280
	16.4.5 Overloading the () operator	280
	16.4.6 Overloading +=	281
16.5	The rule of three	282
	16.5.1 Overriding the assignment operator	282
	16.5.2 Writing a copy constructor	283
	16.5.3 The easy way to abide by the rule of three	284
	16.5.4 Move operators	285
16.6	Completing the Matrix class	285
16.7	Array Programming	286
	16.7.1 Implementing an efficient matrix class	286
	16.7.2 Array programming	287
	16.7.3 Array programming in the option classes	288
	16.7.4 Array programming for the BlackScholesModel	289
	16.7.5 Array programming the Monte Carlo pricer	290
	16.7.6 Performance	290
16.8	Summary	292

17 An Overview of Templates — 295

17.1	Template functions	295
17.2	Template classes	297
17.3	Templates as an alternative to interfaces	299
17.4	Summary	302

18 The Standard Template Library — 303

18.1	typedef	304
18.2	auto	306
18.3	Using iterators with vectors	307
18.4	for loops and containers	309
18.5	The container set	310
18.6	The container vector	311
18.7	The container list	312
18.8	The container initializer_list	315
18.9	The containers map and unordered_map	315
	18.9.1 How a map works	317
	18.9.2 How an unordered_map works	318
18.10	Storing complex types in containers	320
18.11	A mathematical model for multiple stocks	320
18.12	Using the Standard Template Library in FMLib	322
18.13	Summary	327

xiv   Contents

19 Function Objects and Lambda Functions   329

19.1	Function objects	329
19.2	Lambda functions	330
19.3	Function pointers	333
19.4	Sorting with lambda functions	334
19.5	Summary	336

20 Threads   337

20.1	Concurrent programming in C++	338
	20.1.1 Creating threads	338
	20.1.2 Mutual exclusion	339
	20.1.3 Global variables and race conditions	342
	20.1.4 Problems with locking	343
20.2	The command design pattern	346
20.3	Monte Carlo pricing	347
	20.3.1 Random number generation with multiple threads	348
	20.3.2 A multi-threaded pricer	349
	20.3.3 Implementing `Task`	350
	20.3.4 Using the `Executor`	351
	20.3.5 Remarks upon the design	351
20.4	Coordinating threads	352
	20.4.1 The Pipeline pattern	352
	20.4.2 How `Pipeline` is implemented	355
20.5	Summary	358

21 Next Steps   359

21.1	Programming	359
	21.1.1 Libraries	359
	21.1.2 Software development	359
	21.1.3 C++ language features	360
	21.1.4 Other languages	360
21.2	Financial mathematics	361

A Risk-Neutral Pricing   363

A.1	The players in financial markets	363
A.2	Derivatives contracts	366
A.3	Risk-neutral pricing	370
A.4	Modelling stock prices	372
A.5	Monte Carlo pricing	377
A.6	Hedging	379
A.7	Summary	382

| Bibliography | 383 |
| Index | 385 |

Introduction

The aim of this book is teach you C++ from scratch using examples from financial mathematics. It is a streamlined account of the features of C++ that are most useful to a financial mathematician.

Throughout the book we will focus on a key recurring example: How do you price a portfolio of financial derivatives? We will use this example to show

- How to use C++ language in practice

- What kinds of problems banks face

- The skills you need to solve them

These skills include C++ programming skills and mathematical skills but also include testing, debugging, design, and software architecture.

The financial mathematics knowledge needed for this book has been kept to a minimum and is summarised in Appendix A.

Why should you learn C++?

There are many jobs in the finance industry which require sophisticated mathematical skills. One of those roles is being a "quant developer". A quant developer's task is to implement the ideas of financial mathematics in practice to produce practical systems to price, trade, and risk manage complex financial products. This book is aimed at people who already know the mathematics and want to learn the programming skills of a quant developer.

C++ is a programming language. It is just one of many languages that can be used for performing financial calculations. When banks began to develop their trading and risk management platforms, many of them decided that they would write them in C++. As a result, C++ is one of the most sought-after programming skills for quant-developer jobs.

However, it would be wrong to say that C++ is the only programming language worth knowing if you want to be a quant developer. Languages such as C#, Java, MATLAB®, and Python are all heavily used in the financial industry. For someone new to programming, the biggest practical difference between C++ and these other languages is that C++ is much harder to learn!

xvii

xviii *Introduction*

But as a result, C++ skills are also particularly highly valued. Moreover, once you know C++ you will find any of these other languages easy to learn.

In summary, C++ is the language to learn if you want to open up the maximum number of employment possibilities in quant-developer roles. Of course, this probably won't be true forever as technologies do change. So before buying this book, have a search online for your dream job. If C++ is one of the skills required, read on.

Pricing a portfolio

This book will focus throughout on financial examples. As we introduce features of C++ we will show how they can be used to solve real financial problems. Indeed, we will focus on a single important financial problem: How do you compute the price and risk of a portfolio of complex financial products? This is just a simplified version of the real problem faced by a bank of valuing and measuring the risk of their entire position.

It is important to see just how complex this question really is. So let us examine it in more detail.

Any major bank will trade on many different exchanges. Famous examples include the New York Stock Exchange and the London Stock Exchange. However, there are many other less-famous exchanges in cities throughout the world.

Each stock exchange has its own trading rules. Obviously there is some attempt to rationalise things on national and international bases, but contracts and conventions can, and do, vary from market to market.

Of course, one doesn't just trade in stocks. One can also trade in stock derivatives, currencies, currency derivatives, government bonds, municipal bonds, commodities, electricity, etc.

Now let us return again to the problem of calculating the current value of a bank's overall position and the riskiness of that position. Given the complexity of a bank's total position, and the total amount of detail in all the contracts they have entered into, one sees that this is a daunting task. No individual is ever likely to understand every detail of the calculation.

Problem 1. *How do you write software so that no individual has to understand everything that is going on?*

Problem 2. *How do you write software so that a team of hundreds can work on the software at the same time without getting in a mess?*

Problem 3. *How do you write code that is easy for others to understand?*

Another important consideration is that the results of a computer error in financial applications can be catastrophic. If Adobe Acrobat crashes,

Introduction xix

you might swear under your breath. When Barclay's cash machine network crashed, it was headline news in the UK. If your trading algorithm throws away half a billion dollars, it is unlikely that you will get as large a bonus as you were hoping for.

Problem 4. *How do you write code that doesn't contain bugs? How do you ensure that there are no bugs in the code written by a team of hundreds?*

Problem 5. *Given that you probably can't guarantee that there are no bugs, how do you ensure that the effects of a bug are not too harmful?*

Every day, new financial products are invented and creative new financial contracts are devised and sold. It must be possible to rapidly update the software used by a bank to reflect these new contracts. While Microsoft only releases a new version of its operating system periodically, banks update their software on an almost daily basis.

Problem 6. *How do you write code that can be extended easily and rapidly?*

Problem 7. *How do you ensure that no bugs have crept into the latest version of your code, given that you plan to release a new version almost daily?*

Problem 8. *How do you release new code, when all the software has to keep running 24×7?*

If a bank is doing well, their business should be expanding. The bank will be moving into new markets and the data volumes in existing markets will be growing exponentially. Nobody wants investments that have sub-exponential growth!

Problem 9. *How can you ensure that your software will continue to work with exponentially increasing data volumes?*

We can broadly categorise all of these problems as problems of *scalability* and *maintainability*. These problems are the biggest IT problems that banks face. It is these problems that explain why such a large proportion of the employees in the finance sector in fact work in IT rather than finance. And it is problems of this sort that *object-oriented programming* was designed to address.

In this book we will show you how to use the object-oriented features of C++ to solve these problems. In addition we will show you how to use other important programming techniques such as testing, debugging, and design, all of which are essential to building complex financial software.

Why do banks use C++?

We have already discussed why you should learn C++. In case you've forgotten, it is to get a job in finance. But why do banks choose to use C++?

Introduction

Back in 1969, development had started on the computer language C. It quickly caught on because it allows you to write very fast code reasonably easily. Unfortunately that code can be difficult to maintain.

C++ first appeared in 1983. It promised to combine the speed of C with the scalability and maintainability of object-oriented programming techniques. By the 1990s C++ was a mature language which seemed to hold the promise of solving the software problems faced by large financial institutions. This is why banks *started* developing in C++.

The reason that banks have *continued* to develop in C++ is that once you have written a lot of code in one language (and trained your teams to program in that language) it is very expensive to start again in a different language.

These days there are many other languages available that you can use for developing financial software. Some, such as Python, Mathematica, and MATLAB allow you to quickly prototype mathematical ideas but aren't as good for writing large, high-performance systems. Others such as C# and Java are often used for high-performance systems, but just weren't mature enough technologies when banks started their development. Nevertheless, many new institutions such as hedge funds choose to use these more recent languages rather than C++. The people who have invented these newer languages have focussed more heavily on making these languages easy to learn.

One point that is worth emphasising is that many people believe C++ is used because it is the fastest language. It is true that C++ code has the *potential* to match or outperform other code, but it can require great skill and effort to achieve this potential. The reasons for using C++ involve many subtle considerations other than just speed.

In fact, many banks might not choose to use C++ if they started writing their code again today. They continue to use it as the most pragmatic option for their business. In much the same way, it is probably the most pragmatic choice of language for you to choose to learn if you want to work in the banking sector.

Example: Slang

Not all banks choose to use C++ for their quant development. Goldman Sachs makes heavy use of a language called *Slang*. Never heard of it? That's because Goldman Sachs invented the language themselves!

Deciding what language you will write your software in is a major decision. Not only does it affect how easy your system will be to write and maintain, it has more subtle repercussions such as how easy it is to hire people and how good their job prospects are. Do you think it is easy to recruit Slang developers? Do you think Slang developers are highly sought after outside of Goldman Sachs?

Once you have chosen your language and built a system it will cost a lot

Introduction xxi

> to move to a new language. For better or worse, many banks are stuck with C++ and for better or worse Goldman Sachs is stuck with Slang.

The point to emphasise is that C++ is not necessarily "the best language" for financial mathematics. A good quant-developer might decide to use R for statistical analyses, Python for prototyping, C# for developing user interfaces, Excel for their tax returns, and only use C++ to develop code for a legacy trading system. This book focuses on C++, but don't close your eyes to other languages.

How to use this book

The accompanying website for this book[1] contains C++ code you can download and run. This consists of a number of software projects that start from humble beginnings and culminate in a sophisticated financial mathematics library. This gives a concrete demonstration of how one can develop complex software through an incremental process of testing and refactoring. You will find it helpful to download the code for each chapter so you can refer to it easily.

C++ is a computer language. It is almost impossible to learn a language without attempting to speak it! So to get the most out of this book, it is crucial to do as many exercises as possible. The solutions to many of the exercises form an integral part of the software developed in this book. As a result the solutions to many of the exercises can be found by looking ahead at the code for later chapters. The accompanying website for this book contains a solutions guide which shows you where to find the answers to these key exercises.

The order of the chapters has been carefully chosen so that the most useful aspects of C++ are taught first. Therefore, if you are using this book for self study, I recommend that you work through the chapters in order, completing the exercises as you go. If you are using this book as the basis of a lecture course, then the material should be taught sequentially. Chapter 14 and Chapter 15 may be omitted without loss of continuity. Chapter 14 shows how to simulate delta hedging in C++. It is of considerable mathematical interest, but the programming interest is primarily in the exercises. These challenge the reader to design their own polymorphic classes. Chapter 15 discusses debuggers and other development tools. This is very useful practical information, but is perhaps better suited to an interactive class than a formal lecture course.

---

[1] http://www.routledge.com/cw/armstrong

xxii *Introduction*

The book has been designed for a 12-week course at the level of a UK financial mathematics MSc programme. For a shorter course, it would be natural to end at one of the following milestones.

- In Chapter 9 we price a call option using the Monte Carlo method. This gives a practical demonstration of the *procedural programming* skills learned in previous chapters.

- In Chapter 13 we price a portfolio of derivatives using the Monte Carlo method. Since there should be no restrictions on the derivatives that might be in the portfolio, this makes heavy use of *object-oriented* programming.

- In Chapter 18 we extend our model to include markets with multiple stocks. This requires us to use some more interesting data structures and showcases more advanced techniques such as templates and operator overloading.

- In Chapter 20 we show how multiple threads can be used to speed up Monte Carlo calculations. This demonstrates how the C++ techniques described in the book can be combined to build a sophisticated project.

Finally, in Chapter 21 we give some suggestions for further reading and give some general suggestions for possible programming projects to build on the skills you developed by this book.

Acknowledgements

I am indebted to Plamen Turkedjiev, Sohel Rahman, and Alice Sullivan for their many contributions to this book.

Summary

This book provides an introduction to the C++ language and the art of writing high-quality code in that language.

We will illustrate these computer skills with financial examples, specifically the problem of pricing a portfolio of derivatives.

Chapter 1

Getting Started

We start by learning how to write and run a simple example program to compute compound interest. First we will need to install and configure the software required to write C++ programs. Next we will see how to write a simple program.

1.1 Installing your development environment

You will need to install some sort of *development environment* on your computer in order to write C++ programs. A development environment is rather like a word processor, except it allows you to write software rather than documents. Just as with word processors, there are quite a few different development environments you can choose from. We will give some recommendations that have been tested to work with this couse.

1.1.1 For Windows

You will need to download Microsoft Visual Studio Express for Windows Desktop[1]. This software is freely available.

1.1.2 For Unix

If you are using Unix you will need a text editor and the programs g++ and make. If they are not already installed, follow the instructions for your Unix distribution, which you should be able to readily find online. Another program you should install is gdb, which we will discuss in Chapter 15.

1.1.3 For MacOS X

Install Apple's XCode[2] development environment.

---

[1] https://www.visualstudio.com/en-us/products/visual-studio-express-vs.aspx
[2] http://developer.apple.com/technology/xcode.html

1

1.2 Running an example program

Let us start with an example program.

Don't worry what it does for now. The first thing we need to do is work out how to run it.

```cpp
#include <iostream>
#include <cmath>
using namespace std;

int main() {
    // Interesting code starts
    int principal;
    double interestRate;
    int numberOfYears;

    cout << "How much are you investing?\n";
    cin >> principal;
    cout << "What's the annual interest rate (%)?\n";
    cin >> interestRate;
    cout << "How long for (years)?\n";
    cin >> numberOfYears;

    double finalBalance =
        pow(1.0 + interestRate * 0.01, numberOfYears)
        * principal;
    double interest = finalBalance - principal;

    cout << "You will earn ";
    cout << interest;
    cout << "\n";
    /* Interesting code ends */
    return 0;
}
```

You can also find this code on the website of the book[3]. Follow the link for the project "InterestCalculator" for Chapter 1.

---

[3]http://www.routledge.com/cw/armstrong

Getting Started 3

1.3 Compiling and running the code

Running C++ code isn't as easy as you might like.

The reason for this is that C++ is what is called a "compiled language".

The chip inside a computer that does most of the work is called the CPU (central processing unit). It does not know the C++ language or indeed any computer language which is pleasant to program in. The CPU speaks a language called *assembly language,* also known as *machine code* . In machine code, all instructions are coded up as a sequence of numbers. Each number is a code word for some action the CPU should take. Programming in machine code directly is completely unbearable. What is worse, different CPUs may use different versions of assembly language, so you have to rewrite your code for different computers.

To get around this, one programs in "higher level languages" which are written in ways that humans can understand. At some point, the program's instructions need to be converted to machine code.

In an "interpreted" language, the instructions are converted to machine code every time they are executed. MATLAB and Python are examples of interpreted languages.

In a "compiled" language, the instructions are converted to machine code before the program is ever run. This process is called *compilation.* C++ is a compiled language, so you must compile your code before you can run it.

Historically, the advantage of compiled languages was that they run faster. The reason for this is that converting things to machine code takes time. If you do this every time the code is run, it will necessarily run slower. The big disadvantage was that you have to recompile your code if you change the type of computer you want to run it on.

These days, computers are so fast that this advantage is not really relevant any more. Modern languages can be compiled very fast and even use "just in time compilers" that observe how the software is being used by the user and perform optimisations based on this. This is one of the reasons why the claim C++ is faster than languages such as Java and C# is a bit of a myth.

A good development allows you to compile and then run your code at the touch of a button. But before we can do this, we need to get our code into our development environment.

We will now describe the steps you need to go through to compile the example code. Jump to the relevant section for your computer and follow the instructions *to the letter.* Note that a guide to compiling on Macs can be found on the website accompanying this book.

You will need to get everything exactly right. If you cannot get the code to work, there is a zip file called InterestCalculator on the website for this book. This contains working versions of the code that you can use.

1.3.1 Compiling on Windows

- Open Visual Studio.

- Select **File→New→Project ...**

- Select **Empty→Project**

- Enter the Name `InterestCalculator` and press `OK`.

- Note the name of the folder where your project is being saved.

- Notice that to the right of the screen you have an area marked **Solution Explorer** inside which there is a picture of a folder marked **Source Files**. Right click on this and select **Add→New item....**

- Select the option "C++ file" and enter the name `main.cpp` and press `Add`.

- This creates a file called `main.cpp` which we will use to store our code. On the right-hand side of the screen you will see a text editor window where you can edit the code for `main.cpp`.

- Copy and paste the example code from the website[4]into the editor window.

- Select **Project→Interest Calculator Properties...**, then select **Linker→System** and set the **SubSystem** to `Console (/SUBSYSTEM:CONSOLE)` using the drop down.

- Press `OK`

- Press `CTRL + F5` to compile and run your program

This should have worked if you have managed to follow every instruction exactly. If it fails, close Visual Studio, delete all the files in the directory you noted down and try again! But this time be more careful.

Setting up your first-ever project is probably the most fiddly and tedious task you will have to perform in this book.

Why are there so many steps to creating your project?

Firstly a typical C++ project contains a lot of different files, so in practice you don't normally run through such a complex process very often. Most of the steps above are only needed when you create a new project. The two steps that you would be likely to perform repeatedly are:

(i) creating new C++ source files by right clicking on the **Source Files** folder;

(ii) pressing CTRL and F5 to compile and run your program.

---

[4]`http://www.routledge.com/cw/armstrong`

Getting Started 5

The "project" groups together all of your files and allows you to set in one place the configuration options for all your files. This is why it makes sense to have a "project" as well as just the C++ files.

Secondly, you can write different types of programs on Windows. Most programs have Windows user interfaces, but very old fashioned programs have text input through the "console". Console programs are easier to write, but not the default on Windows. So we have to tell Windows that is the kind of program we want. This is why we must set the `SubSystem`.

Danger!

Don't use just F5 or press the "play" button in the toolbar to compile and run your program. Press **CTRL+F5** to run your programs. If you just press F5, the output of your program will disappear the moment the program finishes running, which is confusing at first. In addition, pressing just F5 will execute your program using the Visual Studio debugger, which we do not explain how to use until Chapter 15.

Let us examine all the files that have been created.

If you open Windows Explorer (by pressing the windows key and E) you should be able to browse to where your project has been saved and you will see that a lot of different files have been created.

Most of these are used internally by Visual Studio and so are of no interest. However, the following are interesting.

- `InterestCalculator/main.cpp`. This contains the code we wrote.

- `Debug/InterestCalculator.exe`. This contains the machine code created by the compilation process. You could give someone else with a Windows computer this executable and they could run your program without using Visual Studio.

- `InterestCalculator.sln`. You can double click this to view the project in Visual Studio.

One problem with compiled languages is that when they are running and a problem occurs, the original source code may no longer be available. This makes it very difficult for a program to report where in the code the error actually occurred. This in turn makes it very difficult to debug compiled code.

To help with this, Visual Studio can create executables for you that contain not just machine code but also information about how that machine code corresponds to the original source code. These are called debug executables. By default, Visual Studio will create debug executables. This explains why the executable is in a folder called `Debug`.

Adding in debugging information makes your program bigger and slower.

6   *C++ for Financial Mathematics*

When you finally come to release it to your users you might want to compile what is called a *release* executable without all of this debugging information. To do this you go to the drop down on the toolbar marked **Debug** and select **Release** instead. Now return to the page where you set the properties of your project and use the **SubSystem** option to indicate that the release build is a console application too.

It's a pain that you have to remember to set the properties for both the Release and Debug builds, but it does make some sense that you might want there to be differences in the options used between the two.

For the time being, let's only use the **Debug** executable. Just remember that when you want to see how fast your code really is, you'll want to use the **Release** executable.

1.3.2   Compiling on Unix

The steps to compile and run the code on Unix are as follows:

First create a new directory to store the files for the project.

Next, create a file called **main.cpp** in this directory. Use your text editor to copy in the code for our example program from Section 1.2. You can just copy and paste the code from the website of this book[5].

In the same directory create a file called **Makefile**. Use your text editor to copy in the following text.

```
# Automated Makefile

CC = g++
CFLAGS = -Wall -Werror -D_GLIBCXX_DEBUG -std=c++11 -g
COMPILE = $(CC) $(CFLAGS) -c
OBJFILES := $(patsubst %.cpp,%.o,$(wildcard *.cpp))
PROG_NAME = InterestCalculator

all: myprog

myprog: $(OBJFILES)
    $(CC) -o $(PROG_NAME) $(OBJFILES)

%.o: %.cpp
    $(COMPILE) -o $@ $<

clean:
    rm -f *.o *.html $(PROG_NAME)
```

Note that you needn't try to understand this text, you can just copy and paste it into other projects. When running other programs in this book, the

---

[5]http://www.routledge.com/cw/armstrong

Getting Started 7

only thing you will need to change is the name of the `PROG_NAME` variable. This is just the name of the executable you want to create. One aspect of the file that is worth mentioning is the line

```
CFLAGS = -Wall -Werror -D_GLIBCXX_DEBUG -std=c++11 -g
```

Here we are configuring some compiler options to aid with debugging. For example we are saying that we would like all possible warning messages to be shown (`-Wall`) and that we would like any warning to result in compilation failing (`-Werror`). The flag `-std=c++11` indicates that we are using a relatively recent version of the C++ standard called C++11. We discuss the possible compiler options further in Chapter 15.

Danger!

Note that it is crucial that the tabs on the left of this file are actual tab characters and not spaces!

You can now compile the code by running the following from a command shell

```
make clean all
```

You can then run the code by typing

```
./InterestCalculator
```

The `make` command is a standard Unix tool to make it easier to compile C++ programs. It actually does very little itself other than call another Unix program called `g++` which actually compiles your files into assembly code. The main advantages of make over using `g++` directly are:

- You don't need to type in the names of every single file you want to compile. This is a big help for larger projects.

- If only one file has changed, `make` won't recompile everything.

Running `make clean` gets rid of any code that has been created as part of the compilation. Running `make all` runs an incremental compilation of the files that have changed. Running `make clean all` first gets rid of all the code and then runs a full compilation from scratch.

8　　　　　*C++ for Financial Mathematics*

1.4　Understanding the example code

If you have run the example code on page 6 it will ask you to supply a principal, interest rate and duration of an investment. It then computes the interest that will be accrued.

The mathematics is just compound interest. If the principal is P, the interest rate is i per annum, and the duration of the investment is T, then the interest accrued will be $P(1+i)^T - P$. Note that we're using i here for an interest rate compounded once per year.

The first few lines of code are all what is called "boiler plate" code. This is a term for boring code that you have to write for technical reasons but which doesn't do much. It is called "boiler plate" because boilers often come with a steel plate attached saying who made them and perhaps containing some warnings about how to operate the boiler. Nobody ever reads this, but it has to be there "for legal reasons". The same is true for the first few lines of our code. Its dull, and for the time being we'll skip it.

The first interesting line of code is:

```
// Interesting code starts
```

This is an example of a *comment*. Once you write //, C++ ignores the rest of the line. This allows you to put in helpful comments to guide others through your code.

The last interesting line of code is

```
/* Interesting code ends */
```

C++ also ignores any text sandwiched between the character combinations /* and */. It even ignores new lines. For this reason this is called a multi-line comment.

The first lines that actually do anything are the lines:

```
int principal;
double interestRate;
int numberOfYears;
```

These lines tell the computer to make room in memory for three variables called `principal`, `interestRate`, and `numberOfYears`.

In order to work out how much space it will need to store the data, C++ needs to know what sort of data we will store in these variables. We will store an integer in the variable `principal` and in the variable `numberOfYears`. We will store a real number in the variable `interestRate`.

The phrase "`int principal;`" means: "please make room in memory for a variable called `principal` that will store an integer".[6]

---

[6]If you want to be really picky, it means make room for an integer within a certain range

Getting Started 9

The phrase "double interestRate;" means: "please make room in memory for a variable called interestRate that will store a real number". There are two keywords you can use in C++ to store a real number. You can use float which stores a floating point number to a certain precision, or you can use double which uses twice as much memory but is much more precise. Computer memory is cheaper now than it was in 1969, so float isn't used much any more. The strange name double lives on for backwards compatibility. If you think of double as meaning "real number" you won't get into much trouble.

Notice the semi-colon at the end of the statement "int principal;". Every statement in C++ ends with a semi-colon. You can think of it as the C++ equivalent of a full stop in English. However, C++ is much fussier than the English language—if you forget a single semi-colon, the program won't work.

The line

```
cout << "How  much  are  you  investing?\n";
```

writes the text "How much are you investing?" to the screen and then starts a new line. Anything enclosed in quotation marks is interpreted as text rather than computer code by C++. The special sequence of characters \n means insert a new line.

The line

```
cin >> principal;
```

means read a number typed by the user and store it in the variable principal.

The next few lines behave much the same.

The mathematical heart of our calculation is given by the lines:

```
double finalBalance =
    pow(1.0 + interestRate * 0.01, numberOfYears)
    * principal;
double interest = finalBalance - principal;
```

These lines make room in memory for two variables called finalBalance and interest and immediately assign values to these variables.

The computation of the interest is simple. It is just the finalBalance minus the principal.

The final balance is computed using the formula $P(1 + i)^T$ where P is short for principal, etc. Our code just expands the variable names, replaces multiplication with *, and uses the function pow to raise a number to a given power. In general pow(a,b)$= a^b$.

Notice that C++ doesn't care if your statements go over multiple lines. This is great if you want to write a long formula. The price you have to pay is that you must end all statements with a semi-colon, since C++ doesn't look

---

of values that depends upon whether you have a 32 bit or a 64 bit computer, but that really isn't important right now.

10   *C++ for Financial Mathematics*

at the spacing of your code to guess where one statement ends and another begins. We have indented our code to indicate that the first three lines should be read as a group. You should always space your code carefully to make it easy to read.

Unlike many other languages, C++ doesn't have a special symbol for raising a number to a given power. You might think that the symbol ^ would be a good choice, but C++ in its wisdom uses that for another purpose.

Notice that in mathematics we normally use single letters for variable names such as the formula $P(1 + i)^T$. When writing software, it is usually better to use long variable names. This is because you can then just read the code without having to go somewhere else to find what all the letters stand for.

The next few lines should now be self-explanatory. They print out the answer.

```
cout << "You will earn ";
cout << interest;
cout << "\n";
```

This example shows why C++ makes you specify the line breaks by hand: Sometimes it can be useful to print out only part of a line.

The last few lines are more boiler plate:

```
    return 0;
}
```

If you are curious about what the boiler plate code means, don't worry. All will be explained shortly.

Exercises

One of the most difficult parts of learning C++ is learning how to cope with compiler errors. So this section is very important. I recommend you refer back to it whenever an error happens in your code.

The exercises below show what actually happens if we make some small mistakes in our code.

1.4.1. In the line

```
int principal;
```

remove the semi-colon and then compile the code (recall that on Visual Studio that means press **CTRL F5** and on Unix you should type `make all`). What error message is reported? Do you find this error message helpful?

Getting Started 11

Note that if you are working on Windows, you will have to use the scroll bar to move the text in the **Output** window to the right. Better yet, find the button on the toolbar of the **Output window** to enable word wrap.

Examine the error message. What part of it do you find most helpful? Can you see how the compiler reports which line contains the error? Has it got this right?

Put the semi-colon back, and make sure you can compile and run the code once again.

1.4.2. Repeat the exercise above but removing the) symbol in the calculation of the final balance instead. Make sure you get everything working again before moving onto the next exercise.

1.4.3. Repeat the exercise above but removing the { symbol at the end of the line

```
int main () {
```

Don't panic! Just fix the problem.

1.4.4. Repeat the exercise above, but this time instead of just deleting a character, insert a whole new first line of code that just contains the letter x. Thus the code should start

```
x
#include <iostream>
```

Don't panic! Just fix the problem.

1.4.5. As well as compilation errors, code can contain programming errors that the compiler does not spot. Arguably our example contains one already. What happens if you type "1000$" as the amount you would like to invest? This happens because our code assumes you will only type in numbers.

1.4.6. What error do you get if you completely delete the line `int principal;`? This happens because before you can use a variable in C++ you must tell the compiler what type the variable is.

Tip: Dealing with compilation errors

C++ is very sensitive to tiny punctuation errors. When you get a screen full of errors, don't panic just: **scroll up to the first error, fix that, and try again**.

Once C++ is confused it starts misinterpreting all of your code completely.

So one tiny error can look like a disaster. This is why you should only ever try and fix one error at a time.

You should treat the error messages as the compiler's best guess of what you've got wrong. As the examples above show it can get both the line numbers wrong and the description of the error wrong.

In particular the line number might be slightly below the actual line containing an error, or if you are really unlucky it might think the error is in a different file. The line numbers reported as containing errors actually mark the points where it *realised* that there was an error. This is why they are often below where the real error occurred.

If the compiler tells you that the error is not in your code but in a library it has definitely go it wrong!

You will find it very hard to work out where you have made an error in your code. So always write programs a couple of lines at a time, constantly testing that they compile and run. Then you *know* that any error that appears must be in the lines you've just written.

If you made the mistake of writing an enormous chunk of code and can't work out where the errors are, then delete it all (or comment it out) and put it back piece by piece. Learn from your mistake and start writing code a little bit at a time.

1.5 Configuring the compiler

We have given a detailed recipe for how to configure the compiler to run a simple program. If you are working on Windows, you will see that there are many available options that we are ignoring. Similarly on Unix our Makefile sets some basic compiler options, but other options are available.

We won't worry much about configuring the compiler in this book as it is an advanced topic which only becomes important if you are building large programs that use different libraries or if you want to make sure the code is fully optimised.

We have chosen to configure the compiler so that if errors occur when the program is running, the error messages should be a bit more helpful. We call this "debug mode". Alternatively is also possible to configure the compiler so that it runs as fast as possible. When developing code, or when learning programming for the first time, it is a good idea to prioritise helpful error messages. Once you are happy with your code, you might want to compile it again with different settings so that it runs more quickly. This final version of your code is called the "release" version of your code because it is the version you would release to any users.

1.6   Making decisions

Let's write a new program. Run through a similar process to that given above, but this time let's call our project `ExamCalculator`. Change all the folders accordingly. The code for the `main.cpp` file in this project should look like this:

```cpp
#include <iostream>
#include <cmath>
using namespace std;

int main() {

    int mark;
    cout << "What was your average mark?\n";
    cin >> mark;

    if (mark >= 70) {
        cout << "Congratulations!\n";
        cout << "You got a distinction.\n";
    }
    else if (mark >= 60) {
        cout << "Well done!\n";
        cout << "You got a merit!\n";
    }
    else if (mark >= 50) {
        cout << "You passed.\n";
    }
    else {
        cout << "You failed :-(\n";
    }
    return 0;
}
```

Notice first that all of the boiler plate code is the same as in the last example. This is why I'm calling it boiler plate code. You can just copy it without thinking for the time being.

The more interesting thing to notice is that we are using an `if` statement to make decisions on what the code should do next. The code to be executed in each of the different circumstances is grouped together in sets of curly brackets.

Be careful to notice the round brackets in the condition to be tested.

The `else if` parts of the statement are optional as is the `else`. You can have as many `else if` statements as you like.

14   *C++ for Financial Mathematics*

Strictly speaking, when you only have one statement in a group you can omit the curly brackets. For example, you could finish the code with just

```
} else
    cout << "You failed :-(\n";
return 0;
```

However, it is generally accepted that code with curly brackets throughout is easier to read. So you should always include them.

C++ doesn't care much where you put spaces in your code. However, if you space your code nicely it can be much easier to read. Different authors have different preferences for how to format curly brackets. We will write our curly brackets so that the { appears at the end of a line, the } appears at the beginning of the line, and everything inside the brackets is indented. You will see different conventions in other books. It doesn't matter what convention you choose, but it does make your code more readable if you pick a convention and stick to it.

1.7   Exercises

1.7.1. If a company has revenue r and costs c, then its gross profits, g, are given by $g = r - c$. If it has made a positive gross profit, suppose that a tax of tg must be paid where t is the tax rate. On the other hand, if the company has made a loss, no tax is payable. The net profit is equal to the gross profit with tax paid subtracted.

Write a program called `ProfitCalculator`, which prompts the user for the revenue, the costs, and the tax rate, and then prints out the gross profit and the net profit.

Write two versions, one using single letter variable names and one using long variable names such as `grossProfit`. Note that in C++ variable names can't contain spaces. This is why we use capitals in this way. This way of writing variable names as all one word with capitals in the middle is called *camel case*. The reason is because the word has now got a hump like a camel.

Do you prefer the code with long variable names or single letter names? Which of the two versions would be easier for someone else to understand?

1.8 Summary

We have seen an example C++ program and learned how to compile and run it. We have learned how to work with the error messages produced by the compiler.

Chapter 2

Basic Data Types and Operators

Whenever you use a variable in C++ you must specify the type of data that will be stored in that variable. Moreover, once you have chosen the type of data to be stored in a given variable you can't change it. The jargon phrase is that it is a *statically typed language*. This distinguishes C++ from computer languages which take a more relaxed attitude to specifying the type of data.

The reason that specifying the type of data is important is that ultimately, data is stored on your computer data as strings of 1s and 0s. Integers are encoded as binary numbers. Letters are encoded as numbers written in binary. Floating point numbers are stored in a binary version of scientific notation.

The problem is that if you just look at the 1s and 0s without also knowing the type of the data, the data is completely meaningless. The chip that powers your computer's processing (the CPU or Central Processing Unit) treats data blindly without caring about the type. This means that the CPU is willing to carry out some pretty silly computations if you ask it to. For example, you can ask your CPU to multiply the integer representing the character "w" by the number seven. Probably the only reason you would do this is if your code contains a bug and is accidentally using the same variable for numbers and letters. These kinds of bugs can be found before you ever even run your code if you use a statically typed language.

The C++ language has evolved from the C language. C is also statically typed.

One of the reasons C caught on in the first place is that it is statically typed. The machine-code everyone used before C came along is not statically typed. Programmers quickly found that the C compiler was able to find a lot of bugs automatically by checking types. This meant that they were able to spend more time writing new code and less time fixing bugs.

2.1 Memory terminology

Because data is stored on computers using 1s and 0s, it is natural to store integers using binary.

18   *C++ for Financial Mathematics*

- A single binary digit is called a *bit*. This is just a truncation of the phrase *bi*nary dig*it*.

- 8 binary digits are called a *byte*. This is just a bad joke.

- 1024 bytes make a kilobyte ($1024 = 2^{10}$ so the usual use of "kilo" to mean 1000 is tweaked to work better in binary).

- $1048576 = 2^{20}$ bytes make up a megabyte.

- Numbers representing memory locations are often written in hexadecimal. This is base 16. The 16 characters 0, 1, 2, ..., 9, A, B, C, D, E, and F are used to represent hexadecimal digits. So A is equivalent to the decimal number 10 and F to the decimal number 15. The hexadecimal number 10 can be written in decimal as 16. Four binary digits are equivalent to one hexadecimal digit. For this reason, one hexadecimal digit is called a nibble.

As is already clear, computer programmers have a particularly poor sense of humour.

Old Joke: There are 10 kinds of people. Those who understand binary, and those who don't.

2.2   Basic data types

2.2.1   Integers

You can assign a fixed value to a variable of type `int` using the `=` operator as follows.

```
int numerOfPlatonicSolids = 5;
```

When you type an integer you should not include any commas or spaces. So don't write `1,000,000`, simply write `1000000`.

If you like you can specify an integer using a different numeric base. For example, if you prefer counting in hexadecimal (base 16) you can type `0xFF` instead of `255`. The prefix `0x` means "This number is in hexadecimal".

Confusingly, the prefix `0` on its own means "This number is in base 8". So the code

```
int numberOfPlanets = 08;
```

Basic Data Types and Operators

doesn't compile. Unfortunately, the chances of you remembering this fact are slim! Fortunately, the chances of you making this error are also reasonably low.

When you add two integers, you get an integer, of course. Similarly, if you multiply or subtract them. However...

Danger!

When you divide two integers in C++ you get another integer. C++ simply rounds down if necessary. The C++ code **3/5** evaluates to 0.

C++ actually gives you a number of choices for storing integer data depending on the potential range of values your variable might take. Here are some other data-type specifiers that are available, which all mean an integer of one form or another:

```
short
int
long
long long
unsigned short
unsigned int
unsigned long
unsigned long long
size_t
```

The range of values you can store in each of these data types varies between different C++ compilers. As you might guess, you can store smaller numbers in shorts than you can in long longs. What you can be sure of is that you will be able to store a value between -2^{31} and $2^{31} - 1$ in an **int** variable and a value between -2^{63} and $2^{63} - 1$ in a **long long** variable.

By specifying that a variable is **unsigned** you are saying that it is non-negative. Specifying the sign of a number takes up one bit of memory, so unsigned integers can be twice as big as signed ones.

If you add one to an integer that takes the maximum possible value for that data type, C++ wraps the value round so that it now takes the minimum possible value for that data type. In other words, C++ does its computation in binary and simply discards the top digits if an integer overflows the possible range.

This isn't normally a problem in practice unless you decide to use the **unsigned** data types. For these data types you will find that the calculation $0 - 1$ yields a number greater than 0. This is *extremely* confusing and leads to lots of bugs. For this reason I recommend that you avoid using the **unsigned** data types.

The type **size_t** is used to store the size of data structures stored in memory. Since the maximum amount of memory you can access on a 32-bit

20   *C++ for Financial Mathematics*

computer is 32 bits long, a `size_t` is 32 bits long for 32-bit computers and 64 bits long for 64-bit computers[1].

In summary: Use `int` or `long long` for most purposes; use `size_t` for the size of data structures; don't bother with the other options.

2.2.2 Floating point numbers

Use `double` to store real numbers. Enter numerical values for real numbers by simply entering the decimal expansion with a `.` character for the decimal point.

Notice that C++ considers the numbers `1.0` and `1` to have different data types and so to be different.

If you want, you have a choice of data types for real numbers. You can use a `float`, a `double`, or a `long double`. We won't describe the binary representation of real numbers in detail. However, it is worth knowing the following.

- Most real numbers can only be approximated by a computer. In computer code calculations using real numbers will contain small rounding errors. In particular, when testing your answers you should never test if two decimals are equal, only if they are approximately equal.

- The coding for decimals allows for some special `double` values: `Inf`, `-Inf`, `NaN`, `-0.0`. The first three represent, respectively, ∞, $-\infty$, and not-a-number (for example, $\sqrt{-1}$ is not a real number). The `-0.0` concept has no special purpose, but sometimes calculations give this as the answer. Since you should only ever check if two decimals are approximately equal, the distinction shouldn't cause problems.

2.2.3 Booleans

A `bool` is a variable which can take the value either `true` or `false`. These are special keywords in C++.

`bool` variables are called Boolean variables in polite company. They are named after the English mathematician George Boole.

When you print out a `bool` using `cout <<`, the value `true` is displayed as 1 and the value `false` is displayed as `zero`.

2.2.4 Characters

A `char` is a data type which is intended to store a character.

---

[1] For backwards compatibility, 64-bit computers can run 32-bit programs too. When they are running in 32-bit compatability mode, they behave just as though they were 32-bit computers.

Basic Data Types and Operators 21

Hex	Character	Hex	Character	Hex	Character
20	*space*	40		60	`
21	!	41	A	61	a
22	"	42	B	62	b
23	#	43	C	63	c
24	$	44	D	64	d
25	%	45	E	65	e
26	&	46	F	66	f
27	'	47	G	67	g
28	(	48	H	68	h
29	)	49	I	69	i
2A	*	4A	J	6A	j
2B	+	4B	K	6B	k
2C	,	4C	L	6C	l
2D	-	4D	M	6D	m
2E	.	4E	N	6E	n
2F	/	4F	O	6F	o
30	0	40	P	70	p
31	1	41	Q	71	q
32	2	42	R	72	r
33	3	43	S	73	s
34	4	44	T	74	t
35	5	45	U	75	u
36	6	46	V	76	v
37	7	47	W	77	w
38	8	48	X	78	x
39	9	49	Y	79	y
3A	:	4A	Z	7A	z
3B	;	4B	[	7B	{
3C	<	4C	\	7C	|
3D	=	4D	]	7D	}
3E	>	4E	^	7E	~
3F	?	4F	_	7F	*delete*

TABLE 2.1: ASCII mapping of hexadecimal character codes to visible character codes

In memory a `char` is stored as a number between 0 and 255, or, in hexadecimal, between 00 and `FF`. In other words, a `char` takes up exactly one byte. Some of the mappings between a numeric value and a character are given in Table 2.1.

Unfortunately, C++ was designed to be compatible with C, and C was written by Americans in the 1960s. The authors of C clearly thought that an alphabet of 256 characters would be more than enough for anyone, but for languages like Chinese it is nowhere near enough. Thus a `char` is an appropriate data type to store a character string written in English which may contain some numbers and punctuation marks. However, using the `char` data type is inappropriate for serious programs targeted at an international audience. We will largely ignore these issues in this book, since we are far more interested in working with numbers than working with text.

Interestingly, back when C was being devised, not everyone had computer screens. Some computers just printed their output on paper. For this reason the set of possible `char` values doesn't just consist of letters, numbers, and punctuation, it also contains some instructions for a *teleprinter*. A teleprinter was a primitive printer that behaved more like a typewriter than a modern printer. As a result, there is a special char value `'\n'` which means the instruction "new line". `'\r'` means carriage return. `'\t'` means tab. There is even `'\a'` which means make a ping sound to alert the user that the printout is complete!

You specify a `char` by writing the character in quotes. For example:

```
char theLetterW = 'W';
```

Notice that a `char` represents a single character and not a sequence of characters.

If you are extremely astute you will want to know how you write the characters `'` and `\`, which seem to have special meanings. The answer is that you write `\'` and `\\`. A standard piece of computer jargon is that `\` is an escape character and the process of inserting `\` characters is called escaping.

You can read and write a single characters using `<<` and `>>` with `cin` and `cout` just as we have been doing for numbers.

2.3 Casting

Sometimes you want to convert between one data type and another. This is called *casting*.

Some type conversions that don't result in a significant loss of information happen automatically and without any problem. For example, you can convert an `int` to a `double` without any worries.

Basic Data Types and Operators

```
int anInt = 123;
double aDouble = anInt;
cout << aDouble;
cout << "\n";
```

This prints out 123, as you might expect.

If you have configured the compiler correctly, the code

```
double x = 123.6;
int y = x;
cout << x;
cout << "\n";
```

should not compile.

The C++ specification says that this code is technically legitimate, but that a warning should be issued. The instructions you followed to configure your compiler should have included telling the compiler to fail if there is a single warning. This has been done in Visual Studio by choosing the option to treat warnings as errors. On Unix, our `Makefile` contains some instructions to the compiler in the form of `-Werror` and `-Wall`, which mean much the same thing.

Years of programming experience has shown that automatically converting `double` values to `int` values whether by rounding up or down leads to bugs in your code. Its too easy to inadvertently throw away information. For this reason, the compiler will print a warning when you run this code and you should treat that as a failed compilation.

The warning from my compiler is

```
conversion from 'double' to 'int', possible loss of data
```

Tip: Pay attention to compiler warnings

If the compiler issues a warning about your code you should fix it.

There are two reasons. First, it is almost certainly a bug in your code. Second, if the output of your build is full of warnings, it will be hard to find the first error. (This is important. You might recall that when the compilation fails you should always focus on the first error message).

For this reason you should configure your projects so they fail if an error occurs.

If you are certain that you want to convert a double to an integer by rounding down, here is what you do:

```
double a = 3.141;
int b = (int)a;
cout << b;
```

24   *C++ for Financial Mathematics*

This is called casting. Note the (int), which means convert a into an integer.

Casting a `double` to an `int` potentially loses information, namely the decimal part. Casting an `int` to a `double` isn't a problem, on the other hand. However, casting an `int` to a `float` is risky, since the `float` data type uses binary scientific notation with only a handful of significant figures. So an `int` cannot be represented precisely using a `float`.

Casting a `bool` to an `int` converts `true` to 1 and `false` to 0. Casting an `int` to a `bool` is possible too, but will normally result in a loss of information.

Casting a `char` to an `int` converts a character to the number used to represent that character on your system. This is not a very useful feature unless you are interested in performing text processing. One of the exercises at the end of the chapter suggests you convert text to upper case by using this feature.

Very often, the C++ compiler will automatically convert between one data type and another for you. For example the following code works as you expect without the need for cast statements:

```
int a = 5;
double b = a;
cout << b;
```

It prints out the number 5. However, the following code behaves rather badly.

```
int pi = 3.141;
double r = 4;
double area = 0.5 * pi * r * r;
cout << area;
```

If you run the compiler so that all warnings are treated as errors (as you should), the compiler will spot that there seems to be a typing mistake. The variable `pi` should be a `double` not an `int`. However, the C++ standard technically states that the above code is acceptable and so it is possible to compile it, run it, and get a misleading answer.

Danger!

Always treat warnings as errors.

All though casting often happens automatically, you will sometimes want to perform manual casting. One common case is if you want to divide two integers to get a double. This code behaves unexpectedly:

```
int a = 3;
int b = 5;
double c = a / b;
```

Basic Data Types and Operators
25

```
cout << c;
```

The compiler does perform some automatic casting, but in the wrong place! It guesses that you want the data to be cast as shown:

```
int a = 3;
int b = 5;
double c = (double)(a / b);
cout << c;
```

To understand this, it is best to break it down into two parts: the calculation and then the cast. The code above is equivalent to:

```
int a = 3;
int b = 5;
int divisionResult = 3 / 5;
double c = (double)divisionResult;
cout << c;
```

It should now be clear why the code fails. Two ways you can write the code so that it does what you want are:

```
int a = 3;
int b = 5;
double c = ((double)a) / b;
cout << c;
```

and

```
int a = 3;
int b = 5;
double aAsDouble = (double)a;
double c = aAsDouble / b;
cout << c;
```

The syntax we have used here for casting comes from C and so is called a C-style cast or an old-style cast. In C++ you can use an alternative notation for casting, which is called a C++-style cast or a new-style cast. Here is an example of a C-style cast to an integer:

```
double pi = 3.141;
int piRoundedDown = (int)pi;
cout << piRoundedDown;
```

Here is an example of a C++-style cast:

```
double pi = 3.141;
int piRoundedDown = static_cast<int>(pi);
cout << piRoundedDown;
```

26   *C++ for Financial Mathematics*

In my view, the C++-style syntax is bulky to type and hard to read. However, many authors recommend using C++-style casts. In this book we will use C-style casts, if only to save space in printouts.

2.4 Memory addresses

Another data type which is important in the C language is a memory address, more commonly known as a pointer. Manipulating memory directly is just the kind of thing that CPUs are good at. The aim of the C language was to make it possible to write code that is similar to machine code in terms of performance but is easier for humans to read and write. For this reason, pointers are one of the basic data types in C. Since C++ contains C, pointers are a basic data type in C++ too.

However, our aim is to do mathematics and not to manipulate computer memory directly. For this reason we will try to avoid pointers wherever possible. We will only discuss them seriously in Chapter 11. Most C++ courses introduce pointers very early. As a result new programmers get the idea that they should use pointers a lot in their code too. However, an experienced C++ programmer knows that pointers are only an appropriate tool for solving problems involving direct memory access. We will rarely use them.

Nevertheless, for the sake of completeness we will give a brief outline of the concept now.

If you have a 32-bit computer, you can think of your memory as consisting of up to 2^{32} bytes of data. For 64-bit computers, simply replace the 32 with 64. We'll assume a 32-bit computer for the remainder of the discussion.

Each of these 2^{32} bytes is stored in a memory location. Each possible location is itself labelled with a number. This number is called the memory address and it takes a value between 0 and $2^{32} - 1$. Usually human beings aren't very interested in the precise value of a memory address, so they aren't often written down. But when one does write one out, it's normal to write them in hexadecimal. In hexadecimal we can say that a memory location is a number between 00000000 and FFFFFFFF. This helps to clearly distinguish memory locations from "normal" numbers.

A memory address is stored in C or C++ using a pointer data type. So you know what type of data is stored at that memory address, the pointer data type consists of two parts: the name of a data type that is pointed to and a * character. For example, if we write:

```
double* dPointer = 0x00000000;
```

then we have just declared a variable of type **double*** which means "a pointer to a double" or equivalently "a memory location containing a double". We have specified the precise location that it points to, though it is extremely unusual to do this directly.

The one place where beginners are most likely to encounter a pointer is in so-called C-strings. Look at the following code:

```
const char* speech = "To be or not to be?";
```

This creates a variable called `speech` which is of type `const char*`. The `char*` means it is a pointer to a memory location containing a character. The `const` in front means that you can't use the variable `speech` to change the text, only to read it.

The memory location that speech points to contains the first letter of the speech "T". The next memory location in sequence contains the letter "o". The text of the speech is stored in consecutive memory locations. To mark the end of the speech, the character code 0 is used. This is illustrated in Figure 2.1. In this example, the memory location associated with the text is AB102C02 as this is where the text starts.

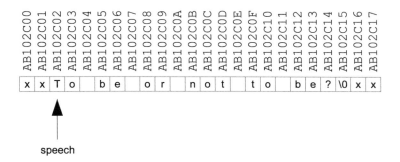

FIGURE 2.1: Text stored in memory and a pointer to the start of the text. The memory addresses are shown in hexadecimal.

Thinking of text in terms of pointers to memory locations is a pretty difficult way to work. It is strongly recommended that you avoid doing this in C++. Instead you should use the data type `string`. To use this you need to have the line

```
#include <string>
```

near the top of the file with the other `#include` statements. You should also have the line

```
using namespace std;
```

as usual. You can then create a variable of type string as follows:

```
string speech = "To be or not to be?";
```

28   *C++ for Financial Mathematics*

If nothing else this is easier to read than the version using `const char*`. Use `string` as your preferred way of storing text.

Technically a `string` is not a basic data type in C++. This is why you have to have write `#include <string>`. By default, only the basic data types are available in C++ unless you have an `#include` statement to indicate that you want to use a more complex data type.

We will talk at length about complex data types in this book. In fact, to some extent this book is about creating your own sophisticated data types in C++. We will create data objects to represent things such as options and market data.

2.5 Operators

2.5.1 The `sizeof` operator

You can find the size (in bytes) of a data type using the `sizeof` operator. For example, the code

```
size_t charSize = sizeof(char);
cout << "A char takes up ";
cout << charSize << " byte\n";
```

can be used to confirm that a `char` takes up one byte as claimed.

2.5.2 Mathematical operations

You can perform various operations on numbers. We have already discussed $+$, $*$, $-$ and $/$.

One other mathematical operator that you probably wouldn't think of is `%`. This is called the modulus operator. It computes the remainder on division. So, for example, `13 % 5` is `3`.

We have also already seen that you can't raise a number to a given power using the `^` operator. Instead you must use the function `pow`.

This function is defined in the library called `cmath`. The statement

```
#include <cmath>
```

at the beginning of all our programs means "we want to be able to use the functions in the `cmath`" library.

There are plenty of other functions in the `cmath` library, such as `sin`, `cos`, and so forth. The syntax for calling a function in C is the same as in mathematics.

```
double x = 2.345;
```

Basic Data Types and Operators

```
double y = sin(x) + cos(1.234);
cout << "The answer is " << y << "\n";
```

If a function requires two arguments, then, just as in mathematics, you separate the arguments with commas.

```
cout << "The 32nd power of 2 is ";
cout << pow(2, 32);
cout << "\n";
```

The available functions in the `cmath` library are listed here: `http://www.cplusplus.com/reference/cmath/`.

2.5.3 Comparison operators

C++ also contains operators for performing comparisons. We have already met >, <, >=, and <=. Also available is == (two equal signs) which is used to test if two numbers are equal and ! = which is used to test if two numbers are different.

Danger!

Use two equal signs == to compare numbers not one equal sign. One equal sign is for *assignment*

2.5.4 Logical operators

The result of a comparison operation is a `bool` representing either `true` or `false`. You can then combine `bool` values using the operators &&, ||, and !, which mean and, or, and not, respectively.

So, for example, `true && false` evaluates to `false`.

2.5.5 Bitwise operators

The operators &, |, ^, ~, << and >> treat integers as simply binary data and perform manipulations on these bits.

- ~ is the bit inversion operator. So ~x contains binary representation of x but with all the 1's and 0's switched.

- << shifts all the bits in a number to the left by a given amount. So x<<3 shifts all the bits 3 steps to the left.

- >> shifts all the bits in a number to the right by a given amount. So x>>3 shifts all the bits 3 steps to the right.

30   *C++ for Financial Mathematics*

- & is the bitwise AND operator. The n-th bit of a&b is equal to 1 if the corresponding bit of a is 1 AND the corresponding bit of b is 1.

- | is the bitwise OR operator. The n-th bit of a|b is equal to 1 if the corresponding bit of a is 1 OR the corresponding bit of b is 1.

- ^ is the bitwise exclusive or operator (XOR). The n-th bit of a^b is equal to 1 if the corresponding bit of a is 1 OR the corresponding bit of b is 1, BUT is 0 IF BOTH ARE 1.

These operators will be of no interest to us in this book, but are very useful if you are working with computer graphics. For example, you can use the NOT operator to generate the negative of a black and white image.

The only thing you have to remember is that you want to use && to mean "and" and not &.

Danger!

The operator ^ does not mean raise to a power, it means XOR.

2.5.6   Combining operators

You can build up complex expressions by combining various operators. However, you should be careful to use brackets to make your desired meaning clear.

The precedence of some operators is well known. You probably know the BODMAS rule for arithmetic that dictates that brackets come first, then division, multiplication, addition, and subtraction. We've underlined the letters that give the BODMAS rule its name.

The precedence of other operators is comparatively obscure. Which has higher precedence, && or ||? It is not obvious. If in doubt, add brackets. I recommend that you *do not* memorise the rules for operator precedence beyond the BODMAS rule. It's better to use brackets for the benefit of everyone who has not memorised them.

2.5.7   Assignment operators

The operators such as * and + that we have been discussing simply compute a value. We have also seen the = operator. This assigns a value to a variable as in the code:

```
int i;
i = 3;
```

As we have already commented, to use a variable in an expression, we have to have said what the type of the variable is at some point. This is why we need to declare the type of the variable before we can use it in assignment statements.

As a convenient shorthand, C++ allows you to combine assignment with arithmetic in a single operator. For example, suppose that you want to add 3 to an integer variable called i then instead of typing i=i+3, you only need to type i+=3.

There is also an assignment operator -= for combining assignment with subtraction, and similarly there are operators *= and /=.

The following code is valid C++

```
int i = 3;
int j = 0;
j = (i += 4);
cout << "The value of i is " << i << "\n";
cout << "The value of j is " << j << "\n";
```

It prints out that the value of i is 7 as is the value of j. The way this code works is that the expression i+=4 adds 4 to the value of i to compute 7. The expression i+=4 is then itself considered to take the value 7, which is the value assigned to j.

Here is a slightly more complex example:

```
int i = 3;
int j = 0;
j = 3 * (i += 4);
cout << "The value of i is " << i << "\n";
cout << "The value of j is " << j << "\n";
```

This time i works out as 7 and j works out as 21.

In general, an assignment expression both *assigns* a value and *has* a value, which can be used for further computations.

Having said this, using the value of assignment expressions in this way makes your code very confusing, so you shouldn't actually do it. Unfortunately, you will sometimes do it by accident. Consider the following code:

```
bool i = false;
bool j = true;
bool areIAndJEqual = (i = j);
cout << "The value of areIAndJEqual is ";
cout << areIAndJEqual << "\n";
cout << "The value of i is ";
cout << i << "\n";
```

The problem here is that the code contains a subtle bug. The expression i=j should probably read i==j and this tiny difference changes the meaning of the code!

The next assignment operator is ++. It adds one to a variable. For example, try the following code:

```
int c = 5;
c++;
cout << c << "\n";
```

It should print out 6. This is why C++ is called C++. It is the next thing after C. Did I mention that computer programmers like weak jokes?

You can also write ++c to increment the variable c. The difference between c++ and ++c is that c++ takes the value of c before the c was incremented and ++c takes the value of c after c was incremented.

In other words, the code:

```
int c = 5;
int d = c++;
cout << "c=" << c << "\n";
cout << "d=" << d << "\n";
```

will give the result that c=6 and d=5. Whereas

```
int c = 5;
int d = ++c;
cout << "c=" << c << "\n";
cout << "d=" << d << "\n";
```

will give the result that c=5 and d=6.

Danger!

Using the assignment operators =, +=, ++ etc. as part of more complex expressions is very confusing. Don't do it.

2.5.8   If statements revisited

Recall that the syntax for an if statement is:

```
if (expression) {
    statements
} else if (expression) {
    statements
} else if (expression) {
    statements
} else {
    statements
}
```

Basic Data Types and Operators 33

The `expression` is allowed to be any basic data type. The value 0 is interpreted as `false` and other values are interpreted as true.

So, for example, the following code snippet prints out that the test has passed.

```
if (-1.743) {
    cout << "Test passed\n";
}
```

The problem occurs when you use = in tests by accident instead of ==. The following is legal C++:

```
int i = 1;
int j = 3;
if (i = j) {
    cout << "i is equal to j\n";
}
```

In the if statement, the code assigns a value of 3 to the variable i and then, observing that 3 is non-zero, prints out the false claim that i is equal to j.

Danger!

Do not use = in `if` statements. Use == instead. Make the compiler treat warnings as errors to prevent this problem.

Exercises

Tip: Getting your code to work

If your code doesn't compile, look at the advice on removing compiler errors in the last chapter.

Some common errors to watch out for in beginner's programs are:

- forgetting the semi-colon character;

- brackets that don't match;

- forgetting either the curly brackets or the round brackets in `if` statements;

- forgetting to enable treating warnings as errors.

34   *C++ for Financial Mathematics*

2.5.1. Write down, or compute, the values of:

- `true || false`
- `(true && false) || true`
- `true && (false || true)`
- `true && false || true`
- `3*5==15 && (7*8==21 || true!=false)`

The moral of this exercise is that you should use plenty of brackets and that you should split complex expressions into small ones.

2.5.2. Create a table of all the different basic data types and the number of bytes that each uses. Use `sizeof` to find the answers. Try out various combinations of `long` and `unsigned`, etc.

2.5.3. Cast `char` values to `int` values to find codes used for the characters 'a', 'z', 'A', 'Z', '0', and '9'. What codes are used for a carriage return, a new line, and a tab?

2.5.4. What value do you get if you subtract the `unsigned int` 5 from the `unsigned int` 3? Try writing the numbers involved in binary to explain what is going on.

2.5.5. Write a program that reads a single `char` of input and then prints out the same character in upper case. To do this, convert the `char` to an integer and use this to work out if it is a lower case letter or not. If it is, convert it to the code for an upper case letter. Then convert this back to an integer.

2.5.6. Write a program where the user enters a decimal number and the code prints out the nearest integer. You should use casting as part of your solution. Write a second version the easy way by using the `cmath` library.

2.5.7. The following code contains several bugs. Fix them.

```cpp
#include <iostream>
#include <cmath>
using namespace std;

int main() {
    cout<<"Type 0 for stone, ";
    cout<<"1 for scissors, 2 for paper\n";
    cout<<"Enter player 1's move\n";
    cin>>player1;
    cout<<"Enter player 2's move\n";
    cin>>player2;
    if (player1=player2) {
```

Basic Data Types and Operators

35

```
        cout << "Its a draw\n";
    } else {
        diff = player1-player2;
        if (diff==-2 || diff==1) {
            cout << "Player 1 won\n";
        } else {
            cout << "Player 2 won\n";
        }
    }
}
```

2.5.8. What does the following program output and why?

```cpp
#include <iostream>
#include <cmath>
using namespace std;

int main() {
    int a = -1;
    unsigned int b = 0;
    if (a<b) {
        cout << "a is smaller\n";
    }
    else {
        cout << "a is bigger\n";
    }
    cout << "Because we're interpreting a to equal ";
    unsigned int castValue = (unsigned int)a;
    cout << castValue;
    cout << "\n";
}
```

2.5.9. Write the bell character to cout. If your computer loudspeaker is working, it should beep at you.

2.6 Summary

We have learned about the useful data types int, double, char, and bool. We have also learned some technical facts about other data types that we will not use often in this book.

We have learned how to cast between data types and the potential pitfalls.

We have learned about various C++ operators, including arithmetic operators, comparison operators, logical operators, and assignment operators.

We have learned the extreme importance of enabling compiler warnings. If you do not do this, you will write bugs that take you an eternity to find.

Chapter 3

Functions

A function in C++ is a piece of code that you can call to perform some task. We have already seen the function `pow` that raises one number to a given power. The `cmath` library (`http://www.cplusplus.com/reference/cmath/`) contains various other mathematical functions.

As we shall see, functions in C++ are used for much more than just computing mathematical formulae. In C++ all software should be divided into functions, each of which performs a particular task. One builds up complex functions by calling simpler functions. For example, you will write a `black-ScholesCallPrice` function which will price a Call option using the Black–Scholes formula (Equation (A.6)). This in turn will call a `normcdf` function to compute the cumulative distribution function of the normal distribution.

On a larger scale, a bank's trading system will contain a function to value the bank's portfolio. This will call functions to: read the current trading position from a database; extract current market data from some source such as Bloomberg or Reuters; price the many different types of traded product. Each of these tasks is itself very complex and is coded by breaking it down into lots of small specialist functions.

In the introduction to this book we discussed the need to write scalable and maintainable programs. One of the key ingredients is the technique of breaking a problem into simpler pieces, each represented by its own function.

3.1 The C++ function syntax

Consider the mathematical function:

$$\texttt{compoundInterest} : \mathbb{R} \times \mathbb{R} \times \mathbb{Z} \longrightarrow \mathbb{R}$$

given by

$$\texttt{compoundInterest}(P, i, n) = P \left(1 + \frac{i}{100} \right)^n - P.$$

Here P is the principal, i is the annual percentage rate, and n is the number of years. To formally specify a mathematical function, you need to provide the domain and the codomain as well as the formula to use.

37

One provides very similar information when writing a C++ function. First one gives the name of the function and describes the type of data it can be applied to and the type of data it returns. Next one specifies the actual formulae needed for the function.

In C++ one would write the `compoundInterest` function as follows.

```
double compoundInterest(double P, double i, int n) {
    double interest = P * pow(1 + 0.01*i, n) - P;
    return interest;
}
```

On the first line we first provide the type of data the function computes. In this case it computes a `double`. Next we give the name of the function. In this case the name is `compoundInterest`. We then list the names of all the parameters and their types. In this example we have said that the name of the function is `compoundInterest`. We have said that it takes 3 parameters named P, i and n. We have said that the variable P is of type `double`. i is also of type `double` and n is an integer.

The actual code for the function is written between a pair of curly brackets. You can write any collection of statements you like inside the curly brackets— perhaps assigning several variables and some `if` statements, for example. When the computation is complete, the code uses the `return` keyword to say what the result of the computation is.

In summary, the syntax for defining a function is:

```
ReturnType functionName( ParameterType1 parameterName1,
                         ParameterType1 parameterName2,
                         . . .
                         ParameterTypeN  parameterNameN ) {
    ... statements ...
}
```

In any code written after this function has been defined, you can just call the function exactly as though it was a built in function like `sin` or `pow`. So for example we can rewrite our first example program as follows:

```
#include <iostream>
#include <cmath>
using namespace std;

double compoundInterest(double P, double i, int n) {
    double interest = P * pow(1 + 0.01*i, n) - P;
    return interest;
}

int main() {
    int principal;
```

```
        double interestRate;
        int numberOfYears;
        cout << "How much are you investing ?\n";
        cin >> principal;
        cout << " What's the annual interest rate(%)?\n";
        cin >> interestRate;
        cout << "How long for ( years )?\n";
        cin >> numberOfYears;
        double interest
            = compoundInterest(principal,
            interestRate,
            numberOfYears);
        cout << "You will earn ";
        cout << interest;
        cout << "\n";
        return 0;
}
```

In this example, we have replaced the computation of the compound interest with a call to the compoundInterest function.

Note that the file contains two function definitions, one for a function called compoundInterest and another for a function called main. In fact our very first program contained a function called main as well, we just didn't discuss it at the time.

Every C++ application for the "console" must contain a function called main. A console application is just a program that reads and writes text to the screen as though it was still 1973. When the user asks to run your program, the operating system will make sure that the function main is called. There are many other kinds of applications, for example, web applications, windows applications, and mobile phone apps. If you're writing a console application, you need a main function. The rules for other kinds of application are different and beyond the scope of this book.

Here are some things to notice in our example:

- The definitions of the functions are written sequentially in the file. We first define compoundInterest completely, then we define main completely.

- The variable names used when we call the function can be completely different from those used in the definition of the function. When we choose the names of parameters and variables in a function, those names have no meaning outside of the function.

- There is a return statement at the end of each function. When a function has computed the desired value, it sends it back to the caller using the return keyword.

40   *C++ for Financial Mathematics*

What does using a function actually buy us? In this case, the code is a little bit longer than it was before, but does precisely the same thing. So what are the advantages?

- Reuse. Every time we want to compute compound interest, we can simply call the `compoundInterest` function. We never have to write it again.

- Division of labour. The person who writes the main body of the code only has to be an expert in asking the user for input and output. They don't need to understand how `compoundInterest` is actually computed to write the main chunk of code. If you scale this up to an iPad user interface to price a knock-out option, you realise why you might need to divide the task between different team members.

- Modularity. We have broken our problem into distinct simpler pieces. Just as a theorem is easier to follow when broken into lemmas, a program is easier to understand when broken down into functions.

- Testability. In Chapter 6 we'll see how using functions makes our code easier to test.

Tip: Once and only once

You should never write the same code twice. If you want to perform the same task twice, write a function that carries out that task.

A good programmer will avoid using cut and paste. If you copy code from one place to another, you are just increasing the number of lines of code in your program and hence increasing the amount of code you will need to debug.

Example 1: Write a function to compute the kinetic energy of a body of given mass and velocity using the formula: $E = \frac{1}{2}mv^2$.

Solution:

```
double kineticEnergy(double mass, double velocity) {
    return 0.5 * mass * velocity * velocity;
}
```

As this example demonstrates, you can return complex expressions as well as the value of a single variable.

One thing to observe is that this function doesn't print anything using `cout`. Instead it quietly returns the answer using **return**. Using `cout` to print things is great for your first few programs, but we now want to start writing functions that operate silently. As an example, consider the built-in function

Functions 41

`sin`. It returns a value but it doesn't print anything. If you think about it, it would be annoying if it printed anything. When you are asked to write a function to compute something, you should write a function that returns the given value rather than a function that prints the given value. If an exercise wants you to print something it will say so explicitly.

3.2 Recursion

Functions can call other functions. For example, our `compoundInterest` function calls the function `pow`.

A more interesting feature is that functions can call themselves. This programming technique is called recursion.

As an example, consider the following recursively defined sequence:

$$x_n = n\, x_{n-1} \qquad n >= 1$$
$$x_0 = 1.$$

Clearly $x_n = n!$.

One can define a function to compute x_n as follows:

```cpp
int factorial(int n) {
    if (n == 0) {
        return 1;
    }
    return n * factorial(n - 1);
}
```

If this function is called with the value $n = 0$ it will return 1, otherwise it will return `n * factorial(n-1)`. By induction it really does compute $n!$.

One other feature of C++ that this example illustrates is that a function can contain arbitrarily complex statements combining if statements and return statements. Indeed, a function can have multiple return statements. Once a `return` statement has been executed, the processing of the function is complete and all further lines of code are ignored.

Recursion can be a powerful programming technique, and some algorithms are easier to write if one uses recursion. While recursion is a popular topic in computer science courses, we will not use recursion much in this book. The exercises should give you some hints as to what is possible using recursion and some of the pitfalls.

Recursion can be used equally effectively to construct paradoxes and bad jokes as well as software. As an example, the name of the GNU software system

42   *C++ for Financial Mathematics*

which is widely used on Unix computers is a *recursive acronym.* GNU stands for "GNU's Not Unix". It doesn't even deserve a groan, does it?

3.3   Libraries

As we have seen, a C++ console application needs a `main` method. However, there are many other kinds of program such as Windows programs, web applications, and so forth. One particularly important kind of program is a *library.* A library is a collection of software routines that is designed to be used inside other computer programs.

In practice, you are more likely to write libraries than you are to write stand-alone applications. For example, your job may be to write routines to price individual types of financial derivatives. It will then be somebody else's job to assemble your libraries with libraries written by other people in order to, say, write a trading application.

As a result we are not actually very interested in the `main` method of our program at all. We will keep a `main` method for the purpose of testing, but you should stop thinking in terms of writing a program. From now on, you should imagine that the software we are writing in this book is a library for pricing and not a stand-alone program. Our aim is to write useful functions rather than to write a single program.

3.4   Declaring and defining functions

In our example code, we defined the function `compoundInterest` before the `main` function. We did this because the `main` function uses the `compound-Interest` function.

In more sophisticated recursive code, you may want to write functions that both call each other. To support this, C++ allows you to *declare* a function and then *define* it later.

When you *declare* a function, you say what its name is and the types of arguments that it requires. When you *define* a function you say what the function actually does.

In all our function definitions so far, we have put the declaration and definition together. But if you want you can separate the declaration and definition.

Here is a stand-alone *declaration* for a compound interest function.

Functions   43

```
double compoundInterest(double P, double i, int n);
```

Notice that this is exactly the same as the first line of the definition of the function except that the curly brackets have been replaced with a semi-colon.

One can then put the *definition* wherever you like in the code. The *definition* is all the code that we saw before:

```
double compoundInterest(double P, double i, int n) {
    double interest = P * pow(1 + 0.01*i, n) - P;
    return interest;
}
```

So the C++ definition actually contains the declaration at the start.

Here's our example program rewritten so that the `main` method is *defined* before the `compoundInterest` function. Note that the `compoundInterest` function is still *declared* first.

```cpp
#include <iostream>
#include <cmath>
using namespace std;

double compoundInterest(double P, double i, int n);

int main() {
    int principal;
    double interestRate;
    int numberOfYears;
    cout << "How much are you investing ?\n";
    cin >> principal;
    cout << " What's the annual interest rate(%)?\n";
    cin >> interestRate;
    cout << "How long for ( years )?\n";
    cin >> numberOfYears;
    double interest
        = compoundInterest(principal,
        interestRate,
        numberOfYears);
    cout << "You will earn ";
    cout << interest;
    cout << "\n";
    return 0;
}

double compoundInterest(double P, double i, int n) {
    double interest = P * pow(1 + 0.01*i, n) - P;
    return interest;
```

44   *C++ for Financial Mathematics*

```
}
```

Technically speaking, you can actually trim down a function declaration to exclude the parameter names. This is a perfectly valid function declaration:

```
double compoundInterest( double, double, int );
```

Notice the analogy with the mathematical statement:

$$\text{compoundInterest} : \mathbb{R} \times \mathbb{R} \times \mathbb{Z} \to \mathbb{R}.$$

The parameter names in a function declaration are ignored by the computer, and so they need not match the parameter names in the definition.

The idea that you can separate declaration and definition occurs a lot in C++. Another example happens with variables. You can declare their type and you can separately assign them a value. Or if you prefer you can declare and define them simultaneously in one statement.

For example:

```
double principal;
```

declares a variable, whereas the line:

```
principal = 1000.0;
```

assigns a value to the variable. The line

```
double principal = 1000.0;
```

performs declaration and assignment in one step.

Tip: Declarations before definitions

When you write a C++ file, put all your function declarations before the first definition.

Since function declarations don't contain any code (and so can't use other functions), the order of the different function declarations isn't important. You should declare the most interesting functions first so that someone reading your file can quickly understand the intention of the code.

Resist the urge to try and reorder functions so that the code compiles without the need for separate declarations.

3.5   Functions that don't return a value

As well as computing a value, a C++ function can perform tasks such as writing text to the screen, sending data down the network, printing a file, and

Functions 45

so forth. This is an important difference between a computer function and the purely mathematical notion of a function.

Very often you want a function to perform a task and don't actually want to compute a value. To do this, you use the special keyword `void` to describe the return type.

For example, the function `printHello` given below simply prints the message "Hello" and then returns.

```
void printHello() {
    cout << "Hello\n";
}
```

Danger!

If it isn't obvious already, in this fragment of code we've omitted mentioning that you need the line `#include <iostream>` and the line `using namespace std;`. We'll often omit `#include` statements and `using namespace std;` statements.

3.6 Specifying default values

Sometimes you may want to specify default values for parameters. For example, here is the declaration of a function that allows you to price a call option on a dividend paying stock, but which assumes the dividend rate is zero if it is not supplied.

```
double computePrice(double strike,
    double timeToMaturity,
    double spot,
    double riskFreeRate,
    double volatility,
    double dividendRate = 0.0);
```

If you have separate declarations and definitions, you should put the default value into the declaration as we have done here.

3.7 Overloading functions

It is possible in C++ to have two different functions that share the same name but which take different numbers or types of parameters.

For example, you might write the following two functions which are both called `average`.

```cpp
double average(double a, double b) {
    return 0.5 * (a + b);
}

double average(double a, double b, double c) {
    return (a + b + c) / 3.0;
}
```

Although these functions have the same name, they take different numbers of parameters so when you type `average(1.0, 2.0, 3.0)` it is possible for the C++ compiler to work out that you must want to call the second function.

Similarly, here are two possible `max` functions, one operates on integers and the other on reals.

```cpp
int max(int a, int b) {
    if (a>b)
        return a;
    return b;
}

double max(double a, double b) {
    if (a>b)
        return a;
    return b;
}
```

Again the compiler can work out which one you are calling. The code `max(1,2)` would call the first version, whereas the code `max(1.0, 2.0)` would call the second version. This is desirable because we are avoiding unnecessary conversions from `int` variables to `double` variables.

Since the identity of a function is determined by both its name and the types of its parameters, this combination is called the *signature* of the function. The idea is that two people are the same if they have the same signature and similarly two functions are the same if they have the same signature.

If when you call a function, there isn't a version with just the right signature available, C++ will perform automatic casting if necessary. For example, you can type `max(1, 2.0)` and it will call the version of the code that treats all parameters as doubles.

3.8 Global and local variables

The functions that we have written so far interact by passing parameters and returning values.

Sometimes you may think it is a good idea to share a variable between functions.

As an example, there is no standard definition for the number π in C++. So you might want to write the following code.

```
const double PI = 3.141592653589793;

double computeArea(int r) {
    double answer = 0.5 * PI * r * r;
    return answer;
}

double computeCircumference(int r) {
    double answer = 2.0 * PI * r;
    return answer;
}
```

The first defines a `double` variable called `PI` with the given numeric value. It uses the `const` keyword to indicate that the value is not allowed to change.

Because `PI` is declared outside of any function it is called a *global* variable. By contrast, the variable `r` in each function is called a *local* variable.

The names you use for local variables within a function have no relationship with the names you use in another function. For example, we have reused the variable name "answer" in two different functions to refer to different quantities. Because these variables are local variables they don't interfere with each other.

The *scope* of a variable refers to the parts of code where that variable can be used. So we say that `PI` has global scope and `r` has local scope. In C++, the curly brackets determine the scope of a variable. If a variable is first mentioned within a set of curly brackets, it can only be referred to within those brackets. Once the execution of the code leaves those brackets, the variable is deleted. We will discuss scope further in Section 4.7.

Tip: Avoid global variables other than constants

Over time, computer programmers have learned that using global variables makes code hard to understand. When writing programs you should try to divide things into small independent pieces. Using global variables prevents

48   *C++ for Financial Mathematics*

these pieces being truly independent. When you change a global variable that is used by some other bit of code you are unaware of, you may accidentally break that code.

Danger!

In Chapter 20 we will learn how to write code which can execute more than one function simultaneously. This is called multi-threaded code. If you use global variables in multi-threaded code you can have a situation where two bits of code are trying to change the same variables at the same time. This is called a *race condition*. To prevent this happening, you need to use a technique called locking. See Chapter 20 for details on how to do this.

3.9   Namespaces

We wrote a function called `average` in the previous section. In some ways this isn't a very good idea because the word average is ambiguous. Do we mean the mean, mode, or median?

This might not seem a big problem. Now we've noticed the ambiguity, we could just rename the function.

However, suppose we are using two libraries written by other teams (or perhaps completely different companies). In one library they have used the word `average` to indicate the `mean`. In another library they use the word `average` to indicate the `median`. These functions are not written by us, so we can't change them. Will this mean the libraries are incompatible?

Furthermore, changing the name of a function isn't always as easy as it sounds. If we have written a library that somebody else is using, we can't change the name of a function without also changing their code. For this reason, you can't normally rename the functions in a library once you have released the library to your customers.

The same problem will occur if two libraries use identically named global variables.

To get round these problems, C++ has a mechanism called *namespaces*. All global variables and functions have an associated namespace which can be used to identify the function more precisely when necessary. Unless you specify the namespace for your functions, they will be put in the *global namespace*.

In a similar way, English people have a first name that you can use to refer to them and a second name that you can use to help resolve ambiguities.

Functions   49

As an example, the global variables `cin` and `cout` that we have been using extensively are actually declared in a namespace called `std`. If you were to create your own variables with the same names, you could still refer to the familiar variables by using the qualified names `std::cin` and `std::cout`.

In fact, unless you explicitly declare that you are using a namespace with the `using namespace` command, you will have to fully qualify the names. This is why all our programs have begun with the line

```
using namespace std;
```

The line above means that we want to use any code in the C++ standard library without the need to qualify the names. One can get rid of this line, but at the expense of needing to fully qualify the variable name as shown below:

```
#include <iostream>
int main() {
    std::cout << "Hello World\n";
}
```

You can put your own functions into a namespace as follows.

```
namespace geometry {

    double computeArea(int r) {
        double answer = 0.5 * PI * r * r;
        return answer;
    }

    double computeCircumference(int r) {
        double answer = 2.0 * PI * r;
        return answer;
    }

}
```

This code creates two functions and puts them both in a namespace called `geometry`.

We won't write our own namespaces in this book simply because doing so would make our code examples a little longer. However, in real code that you expect to use in practice and share with other people, you should always use namespaces to avoid potential naming conflicts.

Exercises

In the questions below write lots of functions (at least one per question) but only one `main` function that should run each of your functions in turn to check that they work. Notice that the questions are only interested in the functions that you write and not in the `main` method that tests them. So there's no need to write an interactive program, just check that your functions work for a few input values.

3.9.1. Write a recursive function to compute the sum of the numbers between 1 and n.

3.9.2. Write a recursive function that takes two integer parameters a and b and prints out all the numbers from a to b.

3.9.3. The n-th Fibonacci number can be defined by $x_n = x_{n-1} + x_{n-2}$ if $n \geq 2$. We define $x_0 = 1$ and $x_1 = 1$. Write a function `fibonacci` that evaluates the n-th Fibonacci number by recursion. How many times is the function `fibonacci` called in order to compute each of x_2, x_3, x_4, and in general, x_n? Don't worry, we will find a far more efficient way to compute the Fibonacci numbers in the next chapter.

3.9.4. A commonly occurring function in financial mathematics is the cumulative normal function defined by:

$$\texttt{normcdf(x)} = N(x) = \frac{1}{\sqrt{2\pi}} \int_{-\infty}^{x} \exp(-t^2/2) \, dt$$

If $x >= 0$ we define:

$$k = 1/(1 + 0.2316419x)$$

a good approximation for $N(x)$ is given by:

$$1 - \frac{1}{\sqrt{2\pi}} \exp(-x^2/2) k(0.319381530 + k(-0.356563782 + k(1.781477937$$
$$+ k(-1.821255978 + 1.330274429k)))).$$

For $x <= 0$ you can use the same formula to evaluate $1 - N(-x)$.

The formula can be derived by choosing the general functional form and then finding the coefficients that give the best fit. For this question, you should just accept the formula on face value.

Write a function called `normcdf` to evaluate the cumulative normal function. Why would N be a bad name for the function?

3.9.5. Is $\sqrt{2\pi}$ recomputed every time your `normcdf` function is used? Use a global variable to improve this.

Functions 51

3.9.6. For each $n \in N$ define a mathematical function h_n as follows

$$h_0(x, a_0) = a_0$$

$$h_n(x, a_0, a_1, a_2, \ldots, a_n) = a_0 + x h_{n-1}(x, , a_1, a_2, \ldots, a_n).$$

We will call these "Horner functions" because they use the Horner method of evaluating a polynomial. Any polynomial in x can be rewritten as

$$h_n(x, a_0, a_1, a_2, \ldots a_n)$$

for appropriate constants a_i. The advantage of using h to evaluate the polynomial is that you don't have to compute high powers of x.

Implement the first few Horner functions in C++. Give all your functions the same overloaded name `hornerFunction`. Use a Horner function to simplify your `normcdf` function.

> **Note:** This question requires some tedious cutting and pasting. Hopefully you will want to rebel against this and find a better way of solving the problem than just cutting and pasting! However, in this case some tedious cutting and pasting is the right thing to do. This is a rare case where you should violate the Once and Only Once principle. The original code for `normcdf` should have been very fast because it contains no complex code like loops (to be covered in Chapter 4). We're introducing `hornerFunction` to improve readability, but we don't want to harm the speed too much. That's why we don't want to be too clever.

3.9.7. Implement the Moro algorithm for the inverse function of the cumulative normal distribution. Call the resulting function `norminv`.

The Moro algorithm proceeds as follows:

Suppose $x \in [0, 1]$. Define $y = x - 0.5$. If $|y| < 0.42$, define $r = y^2$ and approximate `norminv` with the following formula:

$$y \frac{h_3(r, a_0, a_1, a_2, a_3)}{h_4(r, 1.0, b_1, b_2, b_3, b_4)}.$$

We will define the constants a_i and b_i shortly.

Suppose $|y| >= 0.42$. If y is negative let $r = x$. Otherwise let $r = 1 - x$. Define $s = \log(-\log(r))$. Define t by

$$t = h_8(s, c_0, c_1, \ldots, c_8).$$

If $x > 0.5$, `norminv` is approximated by t, otherwise by $-t$.

Here is a table of values for the constants:

```
a0 = 2.50662823884;
a1 = -18.61500062529;
a2 = 41.39119773534;
a3 = -25.44106049637;
b1 = -8.47351093090;
b2 = 23.08336743743;
b3 = -21.06224101826;
b4 = 3.13082909833;
c0 = 0.3374754822726147;
c1 = 0.9761690190917186;
c2 = 0.1607979714918209;
c3 = 0.0276438810333863;
c4 = 0.0038405729373609;
c5 = 0.0003951896511919;
c6 = 0.0000321767881768;
c7 = 0.0000002888167364;
c8 = 0.0000003960315187;
```

3.9.8. Write a function `blackScholesCallPrice`, which takes five parameters (the strike price, time to maturity, spot price, volatility, and risk-free interest rate) and computes the call option price using the Black–Scholes formula. The formula is given in Equation (A.6) in the appendix.

3.9.9. Do you need to know about Moro's algorithm or Horner's method in order to use the `norminv` function? Explain the connection between this and the problems of scalability and maintainability.

3.9.10. How have you tested your code?

3.9.11. Use the online documentation for C++ to find out about the function `erfc`. What do you need to include to use this function? How can you use this function to simplify some of the code we have written in earlier exercises?

3.10 Summary

We have seen how the use of functions allows us to break our code into small pieces which can be reused.

In the exercises, we have written a function, `blackScholesCallPrice`, which relies on the function `normcdf`, which in turn depends upon the function `hornerFunction`. We can build highly sophisticated programs by combining functions. By getting different programmers to write different functions, we can divide knowledge across a team.

We have seen how C++ allows you to write recursive functions.

We have seen that every C++ function has a namespace, a name, and a signature. We have seen how C++ uses namespaces and types to resolve naming conflicts.

We have learned the difference between local and global variables. We have learned how to define `const` variables.

Chapter 4

Flow of Control

We have already used the flow of control statements `if` and `else` extensively. In this chapter we will learn the other control flow statements provided by C++.

4.1 `while` loops

There are three main ways to perform a task repeatedly in C++: `for` loops, `while` loops, and `do-while` loops. The easiest to understand is a `while` loop.

```
void launchRocket() {
    int count = 10;
    while (count>0) {
        cout << count;
        cout << "\n";
        count--;
    }
    cout << "Blast off!\n";
}
```

To see what this code does, just call the `launchRocket` function from the `main` function.

The flow of control is illustrated in the flow chart, Figure 4.1.

The *body* of the while loop is the collection of statements between the curly brackets.

```
        cout << count;
        cout << "\n";
        count--;
```

This code is executed repeatedly until the condition `count>0` ceases to be true. Recall that the instruction `count--` means "subtract one from count". So the `while` loop will count down from 10 to 1 at which point the program moves onto the first line after the while loop.

This is the line:

55

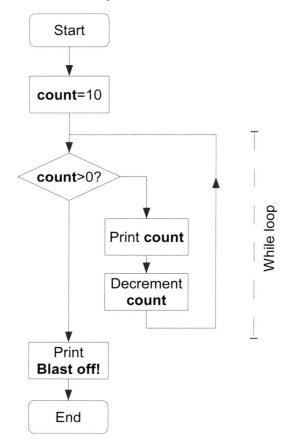

FIGURE 4.1: Flow of control in the launchRocket function.

```
        cout << "Blast off!\n";
```

The general syntax of a while loop is.

```
while (<expression>) {
    <statements>
}
```

Technically speaking, if there is only one statement in the while loop, you can omit the curly brackets. However, it is considered good programming style to always include the curly brackets.

Let us give another example of a while loop. This time we print out the powers of 2 less than 1000.

```
void printPowersOf2() {
```

Flow of Control

```
    int count = 0;
    int currentPower = 1;
    while (currentPower <1000) {
        cout << "2^" << count << "=";
        cout << currentPower;
        cout << "\n";
        currentPower *= 2;
        count++;
    }
}
```

Here we are using the shortcut assignment operator *=.

As one further example, consider the function:

```
void loopForever() {
    while (true) {
        cout << "Still looping\n";
    }
}
```

This program will loop forever. If you run the program, you can stop it by typing **CTRL+C** (on Windows and Unix) or **CMD + .** (on a Mac) in the console window where the program is running.

4.2 do-while loops

These are very similar to while loops except that the test of whether to continue is performed at the end of the loop. For example:

```
void launchRocket_DoWhileVersion() {
    int count = 10;
    do {
        cout << count;
        cout << "\n";
        count--;
    } while (count >= 1);
    cout << "Blast off!\n";
}
```

The general syntax of a do-while loop is

```
do {
    <statements>;
} while (<expression>);
```

58 *C++ for Financial Mathematics*

Because the test is only performed at the end of a `do-while` loop, the body of the loop is always executed at least once.

It is generally accepted that `do-while` loops are hard to read, so they are not used very often. The other looping constructs are preferred.

4.3 `for` loops

It was noticed by the designers of C that most while statements look, in outline, like this:

```
<initialise loop variables>;
while (<test loop variables>) {
    <perform main steps of code>
    <update loop variables>
}
```

To make this repeated design more obvious in code, they introduced the `for` loop. This takes the general form:

```
for (<initialise loop variables>;
 <test loop variables>;
 <update loop variables>) {
    <perform main steps of code>
}
```

To write a `for` loop one simply takes a `while` loop of the form above and moves the code around to match the syntax of a `for` loop.

For example, our code to launch a rocket becomes:

```
for (int i = 10; i > 0; i--) {
    cout << i;
    cout << "\n";
}
cout << "Blast off!\n";
```

To a newcomer to C++, this doesn't look like much of an improvement. The only obvious advantage is that this takes up less lines than the `while` loop equivalent. However, it reads less like English than the `while` loop. If you are new to C and C++ you will probably think the `while` loop is easier to follow.

With familiarity you will grow to prefer the `for` loop. Not because it's easier to understand when you actually read the code, but because you recognise the general shape and so can guess the meaning without looking at the details. Fairly quickly, you'll be able to glance at a `for` loop and immediately

Flow of Control

see what it is doing, whereas with a `while` loop you will need to actually read the code.

Here is the code to count up from 0 to 9 (inclusive). You should commit this code to memory:

```
for (int i = 0; i<10; i++) {
    cout << i;
    cout << "\n";
}
```

Matching this against the general syntax we see that the code to "initialise loop variables" is:

```
int i=0
```

so this code initialises a variable called i and sets its initial value to 0.

The "test loop variables" code is `i<10`. So the body of the loop will be repeated until i is equal to 10. Note that the value 10 will not itself be printed.

The "update loop variables" code is `i++`. So after every loop, i will be incremented.

The "perform main steps of code" is

```
    cout << i;
    cout << "\n";
```

These main steps will be repeated a total of 10 times.

Here is the code to count up from 0 to 90 (inclusive) in steps of 10.

```
for (int i = 0; i < 100; i += 10) {
    cout << i;
    cout << "\n";
}
```

Tip: Start counting at 0

C++ programmers start counting at 0. This is a matter of convention. All the standard data structures in C++ are labelled so that they start with 0.

You might not like it at first, but I strongly recommend that when programming in C++ you start counting from 0 and so use < signs rather than <= signs to compensate.

There are some pretty good reasons for starting counting at 0, because some formulae are easier to write. For example suppose you label the entries of a matrix as follows:

$$\begin{pmatrix} 0 & 1 & 2 & 3 \\ 4 & 5 & 6 & 7 \\ 8 & 9 & 10 & 11 \\ 12 & 13 & 14 & 15 \end{pmatrix}$$

Suppose you label the rows with indices $i \in \{0,1,2,3\}$ and the columns with indices $j \in \{0,1,2,3\}$—then the entry at (i,j) is $4i+j$. It is obvious how this formula generalises to higher dimensional data structures than matrices.

Let's repeat the exercise where we now count from 1. We would label the entries of the matrix as:

$$\begin{pmatrix} 1 & 2 & 3 & 4 \\ 5 & 6 & 7 & 8 \\ 9 & 10 & 11 & 12 \\ 13 & 14 & 15 & 16 \end{pmatrix}$$

We now label the rows with indices $i' \in \{1,2,3,4\}$ and similarly label the columns with $j \in \{1,2,3,4\}$. The entry at (i,j) is now $4(i'-1)+(j'-1)+1 = 4i' + j' - 3$. This is a much less elegant formula than before and it doesn't generalise as straightforwardly to higher dimensions.

Note that there is another kind of `for` loop in C++, which has a simpler syntax but which is compensatingly less flexible. See Chapter 18 for details.

4.4   break, continue, return

In the middle of a loop, you can decide that you want to jump out of the loop and start executing the statements after the loop using the **break** keyword.

For example, this code keeps adding up the numbers typed by the user until they enter a negative number. At this point, the software prints out the total of the positive numbers.

```cpp
cout << "Enter positive numbers followed ";
cout << "by a negative number to quit\n";
int total = 0;
while (true) {
    int next;
    cin >> next;
    if (next <0) {
        break;
    }
    total += next;
}
cout << "The total is " << total << "\n";
```

The **continue** keyword means skip the rest of the code in the body of the

Flow of Control 61

loop, go to the next test statement, and decide whether to continue looping. Thus `continue` means "continue looping" whereas `break` means "break out of the loop".

Here is an example of a continue statement:

```
cout << "Enter positive numbers ";
cout << "Type CTRL+C to quit\n";
int total = 0;
while (true) {
    int next;
    cin >> next;
    if (next<0) {
        continue;
    }
    total += next;
    cout << "Positive total is " << total<< "\n";
}
```

You can also use a return statement in the middle of a loop. This stops all looping and returns the execution to the point where the function was called.

Here is an example of a return from inside a loop:

```
void countdown() {
    int i = 10;
    while (true) {
        if (i == 0) {
            return;
        }
        cout << i << "\n";
        i--;
    }
}
```

Although `break` and `continue` can sometimes make code clearer, they typically make it harder to follow. For this reason newcomers to C++ are best advised to avoid using these statements.

4.5 throw statements

Very often, you will want a function to stop abruptly and indicate that an error has occurred. To do this you use a `throw` statement.

To use throw statements you will need to add the text

```
#include <stdexcept>
```

62   *C++ for Financial Mathematics*

alongside the other `#include` statements at the beginning of your code.

To indicate that an error has occurred, use the following formulation:

```
throw logic_error("You can't do that");
```

You can include pretty much whatever text you like between the quotation marks. Try to provide some clue as to what has caused the problem. When we have discussed string manipulation in Section 7.5 you will be able to create more helpful error messages.

Here is an example of how you might use a `throw` statement:

```
double debitAccount(double balance, double amount) {
    double newAmount = balance - amount;
    if (newAmount<0.0) {
        throw logic_error("No overdraft agreed");
    }
    return newAmount;
}
```

It is much better for a program to `throw` an error and terminate than for it to print out an error message and try to struggle on. This is true for several reasons:

- Nobody is reading the messages! A lot of the code used in finance is running on servers and not on the user's computer. If you write a message to the log file of a server, then the chances are that no one will ever read it.

- It is better to fail obviously than to fail subtly. You don't want someone to think a transaction has gone ahead when it has actually failed.

- If an error is thrown immediately it is easy to see where that error occurred and what was happening when the error occurred. If you look at a log file a day later, you may have no way of working out what was happening.

You might be tempted to try and handle errors in a more sophisticated way than just terminating the program, but the truth is you probably can't. If an error happens in the type of code we are writing, it is probably a programming error and no amount of asking the user to try again later is going to fix the problem. There is nothing for it but to report the error and fix the code.

Dealing with errors in software is an advanced topic in computing. If you are writing an operating system, it is crucial that a bug in a single program doesn't make the entire computer crash. If you are writing a web browser, it is important that a single rogue web page won't make the web browser crash. However, writing web browsers and operating systems are tasks for experts in those fields, not experts in financial mathematics.

If you do wish to handle errors in a more sophisticated way than simply terminating the program you need to use the `try` and `catch` statements. In addition you may want to throw errors of different types to just `logic_error` so that you can distinguish between different types of error. You can find more information in any C++ reference book. We will not need these techniques in this book.

4.6 switch statements

A more complex alternative to an `if` statement is a `switch` statement. Here is an example:

```
void printMessage(int score) {
    switch (score){
    case 0:
    case 1:
    case 2:
    case 3:
    case 4:
        cout << "You have failed.\n";
        break;
    case 5:
    case 6:
        cout << "You have passed.\n";
        break;
    case 7:
        cout << "Merit.\n";
        break;
    case 8:
    case 9:
        cout << "Distinction.\n";
        break;
    default:
        cout << "Invalid score.\n";
        break;
    }
    cout << "Good luck in your future career.\n";
}
```

This program prints out the grade for an exam based on your mark. Marks 0–4 are fails, 5 or 6 is a pass, 7 is a merit and 8 or 9 is a distinction.

The general syntax is:

```
switch (<expression>) {
```

```
    case <possible value>:
        <statements>
    case <possible value>:
        <statements>
    ...
    case <possible value>:
        <statements>
    default:
        <statements>
}
```

The body of the switch statement is the code between the curly brackets. This consists of a list of possible cases and some code to execute. When executing this switch statement, the computer will first evaluate `expression`. It then looks for the first matching case statement and executes all the code from there up to either the first break statement or up to the end of the switch statement.

If there is no match, the computer looks for a `default` statement and evaluates all the code from there to the end of the switch statement.

After executing the required statements from the body of the switch statement, the program moves on to the next statement after the switch statement. In our example this means it will wish you good luck in your career.

Notice that you have to be careful to insert `break` or `return` statements into your `switch` statements to make sure the code doesn't execute more statements than you want it to.

For example, if you remove all the `break` statements from our example, then a score of 0 would be greeted with the response

```
You have failed.
You have passed.
Merit.
Distinction.
Invalid score.
Good luck in your future career.
```

If you dislike the syntax of `switch` statements, you are not alone. Fortunately, they aren't very useful in modern C++ code. In modern C++ one should use the object-oriented programming technique of *polymorphism* instead. We will cover this subject in depth in Chapter 10. If you find yourself tempted to write a `switch` statement, it's a sign that you aren't using object-oriented programming to full advantage.

Tip: Always include a default case

If you must write a `switch` statement, always be sure to include a `default` case. Usually your default case will generate an error by using a `throw` statement. Simply printing an error (as happens in our example) is a bad idea.

4.7   Scope

The flow of control statements above make use of curly brackets. If you define a variable within curly brackets, it can only be used within those brackets. Once the execution of the program leaves the brackets, the variable is deleted. The part of the code in which you can refer to a variable is called the *scope* of the variable.

Here is an example code snippet that computes the sum of the squares of the first 10 integers (starting from 0, of course).

```cpp
int n=0;
int total;
for (int i=0; i<10; i++) {
    int squareValue = i*i;
    total += squareValue;
}
std::cout << n;
// std::cout << squareValue;
// std::cout << i;
```

The variable `squareValue` is only valid within the curly brackets. The variable i is only valid within the for statement itself and within the curly brackets. If you were to try to remove the comment characters `//` on the last two lines, the code would not compile.

The scope of variables is used to control the use of memory in C++. When variables go out of scope, the memory can be safely reused. Often you won't need to think about the memory management in C++. However, there are times when you will want to take control of memory management yourself. We will discuss this in Chapter 11.

4.8   Flow of control in operators

A technical point is that certain operators in C++ actually control the flow of the program in subtle ways.

4.8.1 Short circuit evaluation

The expression `lhs && rhs` evaluates to true if both `lhs` and `rhs` evaluate to true. To improve efficiency, C++ doesn't even bother evaluating the expression `rhs` if `lhs` isn't true. This is usually just what you want to happen. However, if evaluating the expression `rhs` actually calls another function, you may be surprised that this function isn't always called.

Similarly `lhs || rhs` doesn't evaluate `rhs` if it is already known that `lhs` is true.

As an example, try the following code:

```
bool test1 () {
    cout << "In test1\n";
    return false;
}

bool test2 () {
    cout << "In test2\n";
    return true;
}

int main () {
    bool value = test1 () && test2 ();
    cout << "Value " << value << "\n";
    return 0;
}
```

You will see that the function `test2` is never called.

4.8.2 The ternary operator

The operator $*$ as in the expression $a*b$ is called a binary operator because it takes two arguments. The operator `++` on the other hand is called a unary operator.

C++ also contains one operator which takes three arguments and which so is known as *the ternary operator*.

The ternary operator is, in essence, a short cut for writing if statements. The syntax is:

`<test expression> ? <value if true> : <value if false>`

As an example, here is a program that uses the ternary operator to compute the maximum of two numbers:

```
int max (int a, int b) {
    return a>b ? a : b;
}
```

Flow of Control 67

Here is some equivalent code that uses an `if` statement.

```
int max(int a, int b) {
    if (a>b) {
        return a;
    } else {
        return b;
    }
}
```

People who are too lazy to type like the ternary operator, but it results in code which is very hard to understand. Don't use it.

4.8.3   The comma operator

Between the round brackets of a `for` statement one always has three expressions separated by semi-colons. Here is an example:

```
for (int i = 0; i<100; i++) {
    cout << i << "\n";
}
```

You can if you like perform multiple calculations at the end of each loop and multiple initialisations by using the so-called "comma operator". Here's an example that prints out the first 10 powers of 2:

```
for (int count = 0, power = 1;
     count<10;
     count++, power *= 2) {
    cout << power << "\n";
}
```

In this code we maintain two loop variables, `count` and `power`, each of which is updated every iteration.

The formal definition of the comma operator is that the expression `lhs, rhs` evaluates `lhs` and discards the result and then evaluates `rhs`. Thus you can, if you wish, string expressions together using the comma operator as an alternative to using semi-colons in the conventional way.

Don't use the comma operator yourself. It leads to code that is hard to read (as our example demonstrates). We wrote a program earlier (page 56) that printed out the powers of 2 using a `while` loop that was much easier to follow.

Exercises

4.8.1. Use a `while` loop, a `for` loop, and a `do-while` loop to provide different implementations for the `factorial` function.

4.8.2. Given a function $f : \mathbb{R} \to \mathbb{R}$, we can approximate the integral

$$\int_a^b f(x)\,\mathrm{d}x$$

using the *rectangle rule*. Given an integer n, the rectangle rule approximation for this integral is:

$$\frac{1}{n}\sum_{i=0}^{n-1} f\left(a + ih + \frac{1}{2}h\right)$$

where $h = \frac{b-a}{n}$. Write a function that computes the rectangle rule approximation to:

$$\int_a^b \sin(x)\,\mathrm{d}x$$

for given a, b and n.

4.8.3. We would like to compute the integral

$$\int_{-\infty}^x \exp(-t^2/2)\,\mathrm{d}t.$$

By making the substitution $t = x + 1 - \frac{1}{s}$ we can see that this is equal to the integral

$$\int_0^1 \frac{1}{s^2} \exp\left(-\frac{\left(x + 1 - \frac{1}{s}\right)^2}{2}\right)\,\mathrm{d}s.$$

This integral can then be estimated using the rectangle rule. Write a function that uses a for loop to compute this second integral using the rectangle rule.

4.8.4. Use the looping construct of your choice to compute the n-th Fibonacci number. Explain why this is more efficient than the method using recursion considered in Exercise 3.9.3.

4.8.5. Improve the `norminv` function so that it throws an error if asked to evaluate a number outside the range $[0, 1]$.

4.8.6. Improve the `norminv` function so that it takes an extra boolean parameter that indicates whether or not to throw an exception when given an input outside the range 0 and 1. What would be a good default value for this parameter?

Flow of Control 69

4.8.7. Make sure that you can write a `for` loop that prints the numbers from 0 to 9 in increasing order without needing to look anything up in this book. This is the C++ equivalent of being able to count to ten in a foreign language without using a phrase book.

4.9 Summary

We have learned how to control the flow of execution in C++ using the following keywords:

- `for`

- `while`

- `do`

- `break`

- `continue`

- `return`

- `switch`

- `case`

- `default`

- `throw`

By far the most useful flow of control statements are `for`, `while`, and `throw`.

Tip: Which control statement to use

- Use `for` statements for simple loops such as for counting from 0 to 99.

- Use `while` statements for more complex looping logic. Unusual `while` statements are easier to read than unusual `for` statements.

- Use `throw` when errors occur. Don't just print out a message and continue.

Chapter 5

Working with Multiple Files

A realistic C++ project will be divided across many files. This helps you organise your code logically into manageable units. This makes it possible to follow large software projects without having to understand every detail.

5.1 The project FMLib

If you have completed all the exercises in the previous chapters, you should now have developed quite a collection of functions.

In particular, during the exercises we have written the functions `normcdf` (to compute the cumulative density of the normal distribution) and `norminv` (to compute its inverse). We will want to use these functions repeatedly.

This means that it is time to start creating a better structure for our project. We will stop throwing code away and start work on a long-term project called FMLib.

The goal is that FMLib will be a library of useful financial maths functions. The FM stands for financial maths.

We're not the first people to think of writing C++ code for financial mathematics. There are many libraries available that you can already download. However, in this book we will pretend that no libraries exist except those built into the C++ standard. We will build everything else ourselves from scratch.

This is a great way to learn C++, but not an intelligent way to build a real-world trading platform. You should, of course, take full advantage of the work done by other people in writing libraries (and even more importantly the work they have put in to designing, testing, and debugging these libraries). Some libraries that you may want to consider using are Boost, GSL, and Quantlib. Boost is a general purpose C++ library that fills in the many gaps in the C++ standard. GSL is a scientific computing library written in C which contains many useful algorithms. QuantLib is a C++ library for mathematical finance.

At this stage, you will probably find the documentation for these libraries impossible to understand. This is because the documentation assumes considerable expertise in the C++ language.

That is why in the rest of this book we will study how to write our own

71

72   *C++ for Financial Mathematics*

financial maths library FMLib. Once you have mastered this book you should forget about FMLib and learn to use these other libraries instead.

On the website for this book you can find a download called FMLib5. This contains the version of FMLib relevant to Chapter 5. As the book progresses, there are corresponding versions FMLib6, FMLib7, and so forth. These contain increasingly sophisticated code. By the end of the book, we will have developed a fully fledged library.

One aim of this book is to show you how a software project can evolve from simple beginnings to become large and sophisticated. By seeing how a software project can be put together in stages you will learn how to develop your project in manageable pieces. It will also show you how to *refactor* your code. This simply means how to modify your code as you learn from experience to gradually improve your software design.

For the purposes of this chapter, you should download FMLib5. Unzip it and take a look at its contents.

- The file `main.cpp` contains a `main` method for testing.

- The file `matlib.cpp` contains the definitions of various useful financial maths functions. We'll group together many of the standard functions you might expect in MATLAB into this file. For example, `normcdf` and `norminv` are defined in this file.

- The file `stdafx.h` provides a single place to list the libraries we'll want to use throughout the project.

- The file `matlib.h` is a header file containing the declarations of the functions `normcdf` and `norminv`.

As you can see it contains more than one `.cpp` file and some `.h` files. The purpose of this chapter is to explain how C++ code should be arranged using different files in order to develop larger projects and useful libraries. We will use FMLib5 as our example.

5.2   Header files

In C++, you can't directly call functions that are defined in other `.cpp` files. To use a function that is defined in another file, you must first load in a "header file" that contains the declaration of that function using the `#include` command.

In case you have forgotten, we discussed the difference between *declaring* a function and *defining* a function in Section 3.4. Briefly, the declaration says *what* a function does and the definition says *how* it does it.

The rule in C++ is that before you can use a function it *must* have been

Working with Multiple Files 73

declared. So you must always `#include` a file where the function is declared if you want to use a function.

You can think of `.cpp` files as communicating via the `.h` files that they include. If two `.cpp` files both `#include` the same function declarations, then that function only needs to be defined in one of the two `.cpp` files. The same *definition* is shared across multiple files. If two `.cpp` files don't have a `.h` file in common that mentions a particular function, then they can't share that function.

As a matter of convention, whenever we write a `.cpp` file we will typically write another file with the same name, but the extension `.h`. This header file contains function declarations for all the functions we want to be available outside that `.cpp` file.

Some people prefer to use the file name extension `.hpp` for C++ header files and leave `.h` for C header files.

5.3 Creating our project

Let us work through a detailed example of how to create a multi-file project.

Begin by creating a new C++ project called `MyLib`. This is to distinguish it from the model code on the website FMLib. When this section is complete, the projects should in fact be identical.

Our aim is to write a small library that makes available a `normcdf` function and a `norminv` function that compute the cumulative distribution function of the normal distribution and its inverse. We also want to make available a standard constant `PI`.

In a separate file, we want to write a small main function to show how the library can be used.

5.3.1 Creating the first header file

So that other files can use these functions, we will need to declare them in a *header file*. This is a C++ file with the file name extension `.h`.

You should now create a header file called `matlib.h` following the instructions for you development environment below:

- If you are using Visual Studio to create your C++ projects, you should create all `.h` files by right-clicking on the folder **Header Files**.

- If you are using Unix, simply create header files in the same folder as your `.cpp` files but give them the extension `.h`.

- If you are using XCode, press **CMD + N** and select header file.

74   *C++ for Financial Mathematics*

Now edit the file you have just created and type in the first line:

```
#pragma once
```

Then enter the *declarations* for all the functions we want to make available from our library. In our example, the complete code required is:

```
#pragma once

const double PI = 3.14159265358979;

/**
 *   Computes  the  cumulative
 *   distribution  function  of  the
 *   normal  distribution
 */
double normcdf( double x );

/**
 *   Computes  the  inverse  of  normcdf
 */
double norminv( double x );
```

This contains the declarations for all the functions we wish to make available. It contains both the declaration and the definition for `PI`.

In general, you should put the following in your header file:

(i) The first line `#pragma once`. We'll explain why shortly.

(ii) Declarations of functions you wish to make available from your library.

(iii) Declarations of global variables you wish to make available.

(iv) Definitions of constant global variables.

Whereas in your source file you should put:

(i) Definitions of functions declared in the header.

(ii) Definitions of global variables missing from the header.

We will expand on these rules throughout this book as we meet more types of declaration and definition.

The reason why constants are treated differently is that the C++ compiler can "inline" them. This means to replace every use of the constant with the actual value. This gives a slight performance boost. As a consequence, however, the definition needs to be in the header, so that every file knows the correct value to inline.

Working with Multiple Files

5.3.2 Some code that uses the functions

In a source file called `main.cpp` write the following code

```cpp
#include <iostream>
#include "matlib.h"
using namespace std;

int main() {
    cout << "normcdf(1.96)="
         << normcdf(1.96) << "\n";
    cout << "norminv(0.975)="
         << norminv(0.975) << "\n";
    return 0;
}
```

Note that we use `#include` to load in all the definitions in libraries. We use angle brackets when loading in standard libraries and quotation marks for our own libraries.

The reason for this is that our own libraries keep changing all the time, whereas the standard libraries won't. By using angle brackets we are telling the compiler that a given file won't have changed since the last compilation. This allows the compiler to run a bit faster.

At this point, you should attempt to build the code, but you should expect it to fail. After all, we haven't provided any function *definitions* yet.

Try to build the code to see what error message to expect when we forget to write the definition for a function—or indeed what happens when the definition and the declaration don't match precisely.

When I tried to run this code on Visual Studio, I got this error:

```
main.obj : error LNK2019:
unresolved external symbol "double __cdecl
norminv(double)" (?norminv@@YANN@Z) referenced
in function _main
```

This is called a *linker error*. The phrase "unresolved external" is an unhelpful way of saying that either:

(i) you forgot the definition altogether;

(ii) the type information in the definition doesn't exactly match the type information in the declaration;

(iii) you haven't installed a library correctly;

(iv) some more subtle problem has happened. For example, on Visual Studio you need to check whether the `cpp` file is actually listed under "Source Files" in your project.

76   *C++ for Financial Mathematics*

If you examine the error message carefully, you can see that the name of the function (`norminv`) can be found hidden in the text. You just have to be willing to ignore all the parts of the error message you don't understand. It is worth trying to remember what this error message looks like in your development environment so that when a similar linker error occurs you will be able to figure out what the problem is.

The process of turning C++ source files into executables is called "building" the code. This is a three-stage process.

(1) The *pre-processor* performs simple text manipulation on the `cpp` files such as `#include` statements. The statement `#include` for example just means read in a copy of another file so we can use all the declarations from that file.

(2) The resulting `cpp` files are compiled.

(3) The compiled versions of all the files and all the libraries are *linked* together. Each use of a function is *linked* to the place where it is defined.

Function declarations are used during the compile phase and all the function definitions are compiled. However, there is no check that everything glues together coherently until the linker stage. This is why mismatches between declarations and definitions are found at the linker stage.

5.3.3   Write the definitions

Create a file called `matlib.cpp` to contain the function definitions. Start the file with the line `#include matlib.h`.

In general you should have a clear correspondence between your header files and your source files. The exceptions are: It is conventional on Windows to have a file called `stdafx.h` which just contains `#include` statements for the libraries you want to use; you should have a `main.cpp` for testing. We will follow these conventions on all operating systems, not just Windows.

Whenever you write a `.h` file, make sure that the first line of the corresponding `.cpp` file includes the `.h` file. This is an important test that your `.h` file is correct and includes everything that it needs.

In the rest of the file, you should write the necessary code for the function definitions. If you succeeded in writing these in the exercises, you can use your own code. If not, it might be quicker to use the code below.

```
double norminv( double x ) {
    return 1234.0; // TODO fix this
}

double normcdf( double x ) {
    return 1234.0; // TODO fix this
}
```

Working with Multiple Files 77

You should now find that everything builds correctly.

As an experiment, make a change to the type in one of the function definitions as follows.

```
// change double to float
double norminv( float x ) {
    return 1234.0; // TODO fix this
}
```

You should find that you get a linker error, which you can easily fix by reverting to the original code.

5.4 How header files work

5.4.1 The meaning of include

In C++ you must adhere to the following rules about declarations and definitions of functions:

- Every `.cpp` file that uses a function must contain a declaration of that function.

- Every function that is used must be defined in exactly one `.cpp` file or in a library.

On the face of it, this means that we've got a lot of repetitive typing to do. We're going to have to keep declaring every function in every file that uses it. Fortunately C++ has the `#include` command which loads in all the commands from a `.h` file. This way we can load in lots of declarations with one `#include` statement.

There is nothing more to header files than that they save repetitive typing. In principle you could load anything you like using `#include`, but in practice it is only used for loading declarations.

Similarly, in principle you could use a function declared in another file by typing the declaration directly into your own `.cpp` file. There is no technical need to use a header file at all. However, doing this is considered to be bad coding practice. The author of a C++ function uses the header file as a way of indicating which of their functions they are happy for you to use and which they intend to maintain in future.

5.4.2 Pragma once

As we have already said, every header file should start with `#pragma once`. The reason you should start every file with `#pragma once` is that it stops

78   *C++ for Financial Mathematics*

the same file being `#included` twice. In our example, you might get a compiler error saying that you were trying to define `PI` twice if you dropped the `#pragma once` and then `#included` the header twice.

Another rule you should follow is to never have circular dependencies through `#include`. For example, two header files should not `#include` each other.

Technically `#pragma once` isn't part of the C++ standard. However, most compilers understand this command and I recommend you use it.

The phrase `#pragma once` may be hard to remember since there is no such English word as "pragma". It is actually short for the word "pragmatic" which is itself just a fancy word for practical. So `#pragma once` is just a practical fix that isn't really a part of the C++ standard.

If you are a strict adherent to the C++ standard, an alternative way to solve the same problem is to use the following recipe: For each header file pick a unique random name (e.g., G456932745726347569802345237651987624) and then start and end the header file as follows:

```
#ifndef G456932745726347569802345237651987624
#define G456932745726347569802345237651987624
... rest of code ...
#endif
```

You'll see this technique (which is called an *include guard*) used in many C++ libraries. Naturally, many library authors want to write code that is as portable as possible, so they stick to the letter of the C++ standard. That's why they use this more fiddly technique whereas you can get away with the easier `#pragma once`.

5.4.3   Information hiding

You shouldn't declare *all* your functions in your header files. You should only declare the functions you want other people to use.

It is good practice to write lots of small helper functions. However, making these part of your library is confusing for users of your library. Your users will see the `.h` files you write as containing the definition of what your library does. By limiting the amount of information in this file, you make life easier on your users.

It also gives some other advantages. The moment you have released a function as part of a library you can never change it except to fix bugs! This is because someone else might already be using it. The less you give away in your header file, the less potential there is to write functions you have to spend the next ten years supporting.

Example 1: If you misspell the name of a function in a library and then give it to your users, you have to stick with that misspelling forever.

Working with Multiple Files 79

This issue is not so pressing if you are the only person who ever uses your code, but if you work in a team it is absolutely essential.

As a very practical matter, if you have a job writing software and your software goes wrong for some reason, then the chances are high that you will be asked to fix it. As a result, the more people use your code, the more likely you are to get phoned in the middle of the night and asked to debug that code! This means that is essential that you only let people use the code that is of sufficient quality to be shared with the team. This means it should work perfectly, be well documented, and be thoroughly tested. In practice you can't write all your code to this standard, so it is vital to be clear which parts of your code are ready to be shared and which parts are for personal use only. If you don't declare your functions in header files, then that code can't be shared by other files and so *cannot* be used. This means that you can improve your chances of an undisturbed night's sleep if you keep the contents of your header files down to a minimum!

Old Joke: Programming is like sex. One mistake and you have to support it for the rest of your life.

Although simply not including things in the header files is the main way of achieving information-hiding, you can go further with the `static` keyword.

You should try to use the `static` keyword on all variables and functions that are *not* in your header file. For example, one might write the following in a `.cpp` file:

```
static const double SQRT_2_PI = 2.507;
```

This is flawed code in that it is a very low-precision estimate of $\sqrt{2\pi}$. This might not be a big problem for your particular application, but you wouldn't want other people using this in their calculations. That sort of thing results in phone calls in the middle of the night.

By marking this function as `static` we are saying it can only be used in the current source file. This means that we can reuse the name in other source files if desired. It is technically possible (but unlikely) for someone to use a function defined in another file without declaring it in a header file. Using the keyword `static` prevents this trick altogether.

Danger!

If you are a Java programmer, notice that this has nothing to do with the Java keyword `static`.

80   *C++ for Financial Mathematics*

5.4.4   Inline

As a general rule, a good programmer never writes the same code twice. You should write reusable functions. That way you minimise the amount of effort you have to make in testing your code.

Novice programmers often make the mistake of using the cut-and-paste features of their development environment as an alternative to writing functions. This is a very bad idea. It results in code that is hard to read and hard to test. If you detect a mistake in some code that has been manually cut and pasted into 100 other files, you are going to have to track down that error in lots of files.

> **Tip: Write functions. Don't use cut and paste.**

However, from a performance point of view it is ever so slightly faster to run code that has been cut and pasted than it is to write code that has been divided into functions. The reason is that at the level of assembly language, calls to functions are turned into machine code CALL commands. The performance cost of calling a function is therefore equal to the cost of running the code for the function *plus* the cost of a CALL statement.

Most of the time this extra overhead of using functions is absolutely trivial and not worth considering. But every now and again, it is worth thinking about. The solution is not to stop writing functions: it is to tell the compiler to "inline" the functions. This means that you wish to use functions to write your code, but you are telling the compiler to generate the machine code as if you had used cut and paste. The net effect is that the machine code for your function is "copied and pasted" throughout your program by the compiler and so there is no need for CALL statements. In other words, you end up with the faster code you would get by cutting and pasting but without having to sacrifice the benefits of functions.

The `hornerFunction` examples discussed in Exercise 3.9.6 are best inlined. Here is one of the `hornerFunctions` with the `inline` keyword added.

```
static  inline  double  hornerFunction (
        double  x,
        double  a0,
        double  a1)  {
    return  a0  +  x*a1;
}
```

The optimisation can be justified in this special case because we use `hornerFunction` to compute `norminv` and we will be calling `norminv` a great deal in our Monte Carlo simulations. So getting these functions to be as fast as possible is worthwhile.

Working with Multiple Files   81

You might be tempted to make every function `inline` but this may actually make the performance of your program worse. The reason is that all the duplication of machine code that would happen if you did this makes your program much larger and so takes up more memory. This means that your program will take longer to start and will have less memory to play with while it is running.

Modern compilers contain some very sophisticated optimisation routines, so it is usually best not to `inline` anything except for very short functions that you know are performance critical.

Incidentally, notice that `hornerFunction` is also `static` because we don't want it to be part of the library. It is an implementation detail that we don't want our end users to know about.

The reason that we are discussing inline functions in a chapter on dividing your code into multiple files is that inline functions cannot have separate definitions and declarations. This means that if you wish to share an inline function between different files you *must* put the definition into a header file. For all other functions you *must not* put the definition into a header file[1].

This does make some sense. If the compiler is going to copy in the relevant code, it needs to know what that code should be. So that code must be contained either in the `.cpp` file or included from a header file. On the other hand, for non-inlined code the code should only be written once. So that code should only be in one `.cpp` file and should never be included from a header file.

5.5 A complete example

We will provide many versions of FMLib on the website for this book, numbered chapter by chapter. For example, `FMLib5` contains the code needed to understand this chapter.

You should download this library and unzip it into a folder called `FMLib5`. You should then check that you can compile and run the project.

Have a look through the code and observe the following points:

- The `hornerFunction` functions are *not* in the header file. We think users of our library won't want to know about them.

- The `hornerFunction` functions are `static`.

- The constants `a1`, `a2`, etc., used in our calculations are all `static`.

- The `hornerFunction` functions are `inline`.

---

[1] In Chapter 17 we will add an extra item to this rule to cover template functions.

Exercises

5.5.1. Write a new file called `geometry.cpp` which contains a function to compute the area and circumference of a circle given its radius. Ensure that your functions can be called from `main.cpp`. Should your new functions be marked as `static`?

5.5.2. Which file gives a better overview of our matlib functions, `matlib.cpp` or `matlib.h`?

5.5.3. If you mark the function `normcdf` as static in the header file, what are the consequences? In the `cpp` file? In both? What if you don't mark `hornerFunction` as `static`?

5.5.4. Which file do you think should contain the definition for `PI`?

5.6 Summary

We have seen how header files allow us to write projects that use multiple files by putting *declarations* into header files and *definitions* into cpp files. The only exception is for `inline` functions.

Under normal circumstances, two `cpp` files can only share the same function definition if they share the same declaration via a header file.

We have learned the importance of *information-hiding*, which explains why not all declarations should be in header files.

If you have experience programming in other languages, you may wonder why you have to use header files at all. The answer is history. Modern languages do without header files—they provide alternative mechanisms for information hiding that require less typing.

In addition to following the rules of the C++ language, we have also followed some important conventions in this chapter. It is strongly recommended that you follow the following tips when using header files.

Tip: The rules for header files

- Start every .h file with `#pragma once`.

- Don't put a function into a header file unless you have to. Try to hide information.

Working with Multiple Files 83

- Try to remember to use `static` for functions and variables that aren't in the header file.

- For every `.h` file there should be one `.cpp` file that defines everything the `.h` file declares.

- In addition you should have one header called `stdafx.h` that includes the standard libraries you're using. It saves typing to put most of your `#include` statements into one file. It is also possible to use special compiler settings to "pre-compile" this shared header file, which allows you to compile your code more quickly. By convention on Windows this file is called `stdafx.h` and we have decided to use the same convention on all operating systems for simplicity.

- You should have a `main.cpp` file for testing.

- The first line of the `.cpp` files should `#include` the corresponding `.h` file.

- Don't write `using namespace` in a header file. You don't want to force the user of your library to use specific namespaces, it should be their choice. If you were to break this rule and use a namespace in a header file all users of that library will have to use that namespace. This would defeat the entire purpose of the namespace.

Chapter 6

Unit Testing

A revolution occurred in computer programming in the late 1990s (which some banks are still struggling to catch up with). The revolutionary idea was that you should write automated tests for all of your code.

Actually it wasn't such a new idea. Computer programming veterans tell me that in the 1960s and early 1970s, testing your code was commonplace. But somehow lots of development teams forgot about it. It became quite normal to write code without any automated tests. This code was typically very expensive to maintain (with an army of manual testers) and was very prone to bugs.

If you are old enough to remember software from the 1990s, then you might have noticed that software doesn't crash as much as it used to. I put this down, at least in part, to the automated testing boom.

The basic ideas of unit testing are as follows.

(i) Every function (near enough) should have at least one test.

(ii) All tests should be fully automated.

(iii) You should assume that code that is not tested does not work.

(iv) You should keep your tests forever.

Naturally, some manual testing is also required, but this should be much more along the lines of acceptance and usability testing. For example, finding out whether your music download app is *easy to use* would need some manual testing. Testing if your music download app actually *works* should be automated.

If you attempt to write any C++ code that is more than a few lines long, you will probably write a bug. This is partly because you are new to C++, but mostly because you are human and humans make errors constantly. Just as you can be pretty sure that your calculus computation is wrong unless you've double-checked it, you can be pretty sure that C++ code is wrong unless it has a unit test.

One possible way to test your code is to write a **main** method to test the bit you are working on. The problem is that you find yourself constantly recreating tests that you have written before when you need to fix bugs.

The solution is to:

85

86 *C++ for Financial Mathematics*

- write all tests as functions;

- never throw any tests away;

- run all your tests every time you compile your code.

6.1 A testing framework for C++

Since C++ has no built-in support for unit testing, we'll have to write our own. One might argue that it would be more sensible to use a testing framework that somebody else has written. There are lots of choices, such as the Boost Test Library (http://www.boost.org), Google Test (Google it!), cppunit, and many others.

Unfortunately these testing frameworks are designed for people who can already program in C++ and so won't be put off by their odd syntax. This makes them inappropriate to use in an introductory C++ course.

Instead, I've written a very simple testing framework for use in this book. By the end of the book, you will be able to download and use the testing framework of your choice without too much difficulty.

Download FMLib6 and you will see that it contains a header file called testing.h. This file should be incomprehensible to you at the moment as it contains various C++ *macros*.

6.2 Macros

A macro is a special command to the C++ pre-processor which allows you to avoid repeated typing. A classic example is the MAX macro:

```
#define MAX(a,b) (((a)>(b)) ? (a) : (b))
```

This definition means that every time the text MAX(a,b) appears in the code, it should be replaced with (((a)>(b)) ? (a) : (b)). This expression computes the maximum using the ternary operator (see Section 4.8.2 for a reminder of what the ternary operator is). The expression contains lots of brackets which look unnecessary, but which are there because you're allowed to type complex expressions such as like MAX(3*7, 3+5). Having the brackets in place guarantees everything behaves as you expect.

The pre-processor blindly expands macro definitions. This means that you can write "fake functions" like MAX that are really just shorthand ways of writing out a repetitive chunk of code.

Unit Testing 87

Generally speaking, macros are a complete waste of time. You can write reusable code using functions and then use the `inline` keyword to ensure that it is just as fast as the macro code.

However, just occasionally, macros have their uses, because they give you access to special variables like the name of the current file or the current line number. These special features are useful when writing developer tools (such as a testing framework). Unless you are writing such a framework, don't learn how to write macros.

6.3 The macros in `testing.h`

6.3.1 The `ASSERT` macro

```
ASSERT( test );
```

This checks whether a `bool` valued `test` is `true` and throws an error if it is not true.

To speed up performance of a real system, all `ASSERT` checks are skipped when running the release build.

This needs to be a macro because it prints out the line where the assertion failed. You can't do this with functions.

Here is an example of using `ASSERT` to verify that the input arguments to a function are valid. The function below will fail immediately if you attempt to take the square root of a negative number.

```
double safeSqrt(double x) {
        ASSERT(x >= 0);
        return sqrt(x);
}
```

6.3.2 The `ASSERT_APPROX_EQUAL` macro

When writing numerical methods, you don't often want to check if two numbers are exactly equal. You normally want to check if they are approximately equal. This is because numerical methods are only approximate and calculations always involve rounding errors.

`ASSERT_APPROX_EQUAL(x,y,tol)` throws an error if x and y are not within tol of each other. Here is an example of `ASSERT_APPROX_EQUAL` being used to test the output of `norminv`

```
static void testNormInv() {
    ASSERT_APPROX_EQUAL(norminv(0.975), 1.96, 0.01 );
```

88   *C++ for Financial Mathematics*

```
}
```

This example test uses the well-known fact that approximately 95% of the probability density of a normal distribution lies between ± 1.96 standard deviations of the mean. So the 2.5% and 97.5% percentiles of the standard normal distribution are -1.96 and 1.96, respectively. Our test is simply confirming that norminv$(0.975) \approx 0.975$. We will discuss the important topic of how you can still test your code, even when you don't know the correct answers, in Section 6.6.

6.3.3 The `INFO` macro

Using `std::cout` doesn't really make much sense unless you are writing a console application. Real programs often write to a log file which can be examined in the event of a failure, but which are normally ignored on a day-to-day basis.

The `INFO` macro provides a convenient way to print out a message together with the file name and line number where the message was printed. By modifying the macro definition you could easily change how this logging was performed without changing all your code that actually performs logging.

`INFO` also has the minor advantage that you don't have to remember the `\n` on the end of each line.

```
double priceOptionByMonteCarlo( int numScenarios ) {
    if (numScenarios>1000000) {
        INFO(
        "Embarking upon a calculation with "
        <<numScenarios<<
        " scenarios");
    }
    ... /* length calculation goes here */ ...
}
```

Note the way the INFO macro allows you to use `<<`.

In practice you should use a more advanced logging framework. See Chapter 15.

6.3.4 The `DEBUG_PRINT` macro

This behaves exactly like the `INFO` macro except you can control when it prints out its message.

It will print out a message only if

- You are running the debug build.

- You have enabled debug by calling `setDebugEnabled(true)`.

Unit Testing 89

The advantage of this is that your release code won't be slowed down by lots of `DEBUG_PRINT` messages. In addition, you will rapidly find that the `DEBUG_PRINT` messages from code you are not currently working on are annoying. Rather than delete all the messages, simply call `setDebugEnabled(false)` to silence them.

Here's an example of how you might use it:

```
double max(double a, double b) {
    DEBUG_PRINT("Entering max(" << a << ", " << b << ")");
    double ret = a>b ? a : b;
    DEBUG_PRINT("Returning " << ret);
    return ret;
}
```

Serious logging frameworks give you quite fine control over which functions and files have logging enabled. They also give you a user interface so you can interactively enable and disable logging.

Our `DEBUG_PRINT` macro is just a step in the right direction.

6.3.5 The TEST macro

This macro prints out the fact that is about to run a test, runs the test, and then prints out whether the test passed or failed.

Here's an example of how it is used:

```
void testMatlib() {
    TEST( testNormInv );
    TEST( testNormCdf );
}
```

This code simply runs the `testNormInv` function and then runs the `testNormCdf` function.

Although the statement `TEST( testNormInv )` looks a bit like a function call, it is not strictly a function call. This is typical of macros. They look misleadingly like functions but behave in subtly different ways. This is why one normally avoids using them, because it can easily lead to confusion.

6.4 Using `testing.h`

To use the testing framework, you should `#include "testing.h"` in all your `.cpp` files.

For each test that you want to perform, write a function whose name begins `test`. Here is an example:

```
#include "testing.h"

... /*non testing code here*/ ...

static void testNormCdf() {
    ASSERT_APPROX_EQUAL( normcdf( 1.96 ), 0.975, 0.001 );
}

static void testNormInv() {
    ASSERT_APPROX_EQUAL( norminv( 0.975 ), 1.96, 0.01 );
}
```

In each .cpp file you should also write a single function which calls all the
other test functions in turn. Use the TEST macro to do this. This will ensure
you always get output telling you which function is currently running.

You should name this function after the .cpp file, define it in the .cpp file,
and you should declare it in the corresponding header file.

For example, in matlib.cpp we have

```
void testMatlib() {
    TEST( testNormInv );
    TEST( testNormCdf );
}
```

And in matlib.h we have

```
void testMatlib();
```

Finally, in your main method you should call all the test functions defined
in the header files.

```
int main() {
    testMatlib();
    ... /* run other tests */ ...
}
```

When you are trying to understand how your functions are behaving, you
can insert DEBUG_PRINT statements to help you follow what is going on. They
won't be called until you call setDebugEnabled(true). Here is how you
would temporarily enable debug messages when running testNormInv.

```
void testMatlib() {
    // switch on the DEBUG_PRINT statements
    setDebugEnabled(true);
    TEST( testNormInv );
    setDebugEnabled(false);
    // switch them off again
```

```
      TEST( testNormCdf );
}
```

6.5   What have we gained?

- We no longer need to keep writing `main` methods. It is no longer necessary to create new projects to answer exercises in this book. Instead we can simply write a new unit test for each exercise and have a single project with a single main method.

- When trying out our code we don't have to keep changing the main method to check that our code is correct.

- We have a record of all the tests performed.

- Whenever we change our code we can retest immediately.

- We know that our code always works! (So long as we have enough tests).

- We have useful `DEBUG_PRINT` statements that will help us figure out what is going on if we find a bug in future.

These advantages are enormous. So much so that many programmers advocate *test-driven development*. This phrase was introduced by highly influential software developer Kent Beck as part of a general set of programming principles called *extreme programming* [1].

The idea of test-driven development is simple: You should write the test before you write the code. Why is this a good idea?

Firstly it forces you to write the test. Your motivation to write a test will drop once you have a function that already appears to work. Secondly it will force you to think about what problem you are actually trying to solve before you go and solve it. Third it will prevent you writing untestable code. Fourth it tests your tests. Its quite easy to write tests that don't actually test anything, so its good to see that they fail until you've finished writing the code.

Writing tests is a significant skill and one of the most important skills that you should develop by reading this book. A programmer who has learned the benefits of writing unit tests and who won't write code without them is sometimes called *test infected*. It sounds like an insult, but it's a compliment. I hope you become test infected.

Type safety and testing

From a modern perspective, a significant criticism of the C++ language is that it claims to be designed to make large projects easy to write, but it contains no built-in support for testing whatsoever! From a modern perspective this is simply astonishing. Fortunately, there are now plenty of good testing frameworks available to plug the gap.

What C++ has always done well is to make sure that your code doesn't contain any gross errors where you accidentally treat a memory location containing a string as though it contains a double. C++ does this by getting the compiler to check that all the type definitions in your code match up. However, there is a great deal more to getting code to work than ensuring that the type definitions are correct.

Many modern languages don't bother with type safety at all. This is because if your code is tested fully with unit tests, then there won't be any type safety problems. In other words, if you have good unit tests, you won't need to rely on the compiler testing your code for you.

6.6 Testing `normcdf`

The biggest challenge in testing your code isn't writing the C++ code for the tests. The challenge is thinking of the tests to run. It demands some creativity to think of good tests.

Example 1: How should you test the function normcdf, which is intended to compute the cumulative distribution function of the normal distribution?

Solution: Here are some tests you could run on normcdf:

1. Given any value x (say $x = 0.3$) it should be true that normcdf$(x) > 0$ and normcdf$(x) < 1$. These properties hold for any cumulative distribution function.

2. For very small values of x, normcdf(x) should be close to 0.

3. For very large values of x, normcdf(x) should be close to 1.

4. normcdf should be increasing

5. normcdf(x) should be equal to $1 - \text{normcdf}(-x)$ by the symmetry of the normal distribution.

Unit Testing 93

6. normcdf(0) should be equal to 0.5.

7. normcdf(norminv(x)) should equal x.

8. normcdf(1.96) $\approx$ 0.975. This is just a well-known fact: it is worth remembering as it enables you to quickly compute 95% confidence intervals.

9. normcdf(x) $= \int_{-\infty}^{x} \frac{1}{\sqrt{2\pi}} \exp\left(-\frac{x^2}{2}\right) dx$. So it should be possible to test the value of normcdf by computing the integral directly using numerical integration.

Here is the code that runs all these tests but the last:

```
static void testNormCdf() {
  // test bounds
  ASSERT(normcdf(0.3)>0);
  ASSERT(normcdf(0.3)<1);
  // test extreme values
  ASSERT_APPROX_EQUAL(normcdf(-1e10), 0, 0.001);
  ASSERT_APPROX_EQUAL(normcdf(1e10), 1.0, 0.001);
  // test increasing
  ASSERT(normcdf(0.3)<normcdf(0.5));
  // test symmetry
  ASSERT_APPROX_EQUAL(normcdf(0.3),
    1 - normcdf(-0.3), 0.0001);
  ASSERT_APPROX_EQUAL(normcdf(0.0), 0.5, 0.0001);
  // test inverse
  ASSERT_APPROX_EQUAL(normcdf(norminv(0.3)),
    0.3, 0.0001);
  // test well known value
  ASSERT_APPROX_EQUAL(normcdf(1.96), 0.975, 0.001);
}
```

As this demonstrates, writing test code is usually much easier than thinking of tests in the first place. Nevertheless, some tests (e.g., computing `normcdf` by integration) can require quite a bit of work. However, if you are planning to invest real money on the basis of the answers your code produces, or if you are planning to publish a journal article based on your calculations, you should always make sure that you test them as thoroughly as possible.

If you look at the testing code, you will see that we have only tested a few specific cases of the claim normcdf(x) > 0. This is all that is needed. Our tests give us a great deal of confidence in our code even though we have by no means proved it worked in all cases. When you write a test you should be trying to write a quick simple check and not a mathematical proof.

These different ideas for how to test normcdf can be generalised to test many other functions. The last option is the most obvious but is also, in some

94   *C++ for Financial Mathematics*

senses, cheating. If you were the first person in history to write a normcdf function, you wouldn't yet know what the value of normcdf(1.96) should be.

When you come to write complex code, to price some exotic derivatives, for example, you probably won't be able to look up the correct answer. So you will have to come up with rather more cunning tests. Here are some general ideas.

- Are there any mathematical properties that should hold for your answer? Then test that they hold.

- Can you think of alternative ways of calculating the answer? Then implement both methods.

- Does your problem simplify for extreme parameter values? Then check what happens for these extreme values.

- Can you break the calculation down into separate functions which you can test individually? Then do so.

Exercises

6.6.1. Write a unit test for your functions that compute the area and circumference of a circle. Make sure they are actually running.

6.6.2. Insert DEBUG_PRINT statements into the functions to compute the area and circumference of a circle. Check that you can enable the DEBUG_PRINT statements.

6.6.3. The unit test function testNormcdf is marked as static void. Why? The function testMatlib is marked as void but not static. Why not?

6.6.4. Write a function to compute the price of a put option using the Black–Scholes formula, Equation (A.7). Write a test for this function.

6.7 Summary

We've seen how to use the testing.h testing framework. In a production system one would probably use something more sophisticated.

When writing C++, write lots of small functions with tests for each. Constantly compile and run your code to make sure you spot your errors as soon as you make them.

Unit Testing 95

You should try test-driven development and will hopefully become test infected.

Throughout this book we will see many techniques that you can use in order to test your code.

Chapter 7

Using C++ Classes

The types `double` and `int`, etc., are too restrictive. What about complex numbers, strings, matrices ...? To make significant progress with C++ we will need to use more sophisticated data types. Indeed, as you will discover during this chapter, with just a few more data types we will be able to write extremely sophisticated programs. For example, in this chapter we will write the code to produce a very attractive pie chart. As exercises you will be able to write your own functions to produce line plots and histograms.

To use the new data types we will want some `#include` statements. The correct `#include` statements are:

- `#include <string>` to use strings.

- `#include <sstream>` to use strings efficiently.

- `#include <vector>` to work with vectors.

- `#include <fstream>` to work with files.

- Matrices? There is no built-in type for matrices in C++. You can use a library like Eigen, Boost, or Quantlib. Or you can write your own matrix type. We'll do this in Chapter 16.

We'll assume that the relevant `#include` statements have all been added to the file `stdafx.h` so that we can use all the libraries easily. We'll also assume that you've written `using namespace std;` at the top of all your `cpp` files.

We'll make this kind of assumption without comment from now on.

7.1 Vectors

A vector in `C++` is the name given to a data structure containing an numbered list of elements of a given data type. A vector of `double` data corresponds to the mathematical notion of a vector. The name of this type in C++ is written as `vector<double>`.

We'll now see some examples of how you work with vectors.

97

```
// create a vector
vector<double> myVector;

// add three elements to the end
myVector.push_back( 12.0 );
myVector.push_back( 13.0 );
myVector.push_back( 14.0 );

// read the first, second and third elements
cout << myVector[0] <<"\n";
cout << myVector[1] <<"\n";
cout << myVector[2] <<"\n";
```

We see from this example that you use the function `.push_back` to add data to the end of a vector. Notice the way we use a `.` before the function name. This is an example of a *member function* of the vector. It is a function that "belongs" to the vector itself and performs functionality specific to that vector. This is typical of *object-oriented programming*. Objects gather together data and functions for working with that data. So a `vector<double>` is a type of object that stores numbers and provides the function `push_back`.

You always use a `.` in this way when calling a member function. You can think of the `.` as meaning much the same as "apostrophe s" in English. We're calling myVector's `push_back` function.

To read the data from the vector, we're using the index notation `myVector[i]` to access the element at index `i`. Notice that in C++ a vector's indices start at 0. This is one of the reasons why C++ programmers count from 0 when writing `for` loops.

We can also use the index notation `myVector[i]` to change the elements of a vector as shown below.

```
// change the values of a vector
myVector[0] = 0.1;
myVector[1] = 0.2;
myVector[2] = 0.3;
```

You can use index notation to start writing more sophisticated programs with `for` loops. Here's a simple example to print out the elements of a vector using a `for` loop.

```
// loop through a vector
int n = myVector.size();
for (int i=0; i<n; i++) {
    cout << myVector[i] <<"\n";
}
```

Again, notice that we start counting from zero and stop counting using a less-than sign. If you developed this habit when writing your C++ code you will avoid a lot of bugs!

> **Tip: C++ programmers count from 0**
>
> Because the first index of a vector is 0, you should always count from 0 in your `for` loops.

We've just seen another member function of `vector<double>` in this last example. We're using the member function `size()` to compute the size of a vector.

You may be tempted to write the code:

```
for (int i=0; i<myVector.size(); i++) {
...
}
```

rather than compute the size on a separate line. This is technically correct C++, but some compilers (for example Visual Studio) issue a warning if you do this. The reason is that the type of `vector.size()` is an unsigned integer and the compiler is concerned that you are comparing signed and unsigned data types. The fix to this problem is simply to write:

```
for (int i=0; i<(int)myVector.size(); i++) {
...
}
```

By casting the size to a signed `int`, we stop the compiler from fretting. Similarly, if one attempts to compile using 64-bit Windows, you need to cast the return of `size()` before trying to store it in an int because the compiler is concerned that you are trying to store a potentially 64-bit number using only 32 bits. In practice, we will not create vectors with 2^{31} or more elements! As an alternative we could use variables of type `size_t` ourselves. As we have discussed, working with unsigned integers can cause confusion, which is why we prefer the option of casting to an `int`.

Let's see some more examples of what we can do with a vector.

- We can initialise one to be of a certain size and with certain fixed values.

- We can initialise a vector to be a copy of another vector.

- We can replace the contents of one vector with the contents of another vector.

These actions are demonstrated in the code below:

```
// Create a vector of length 10
// consisting entirely of 3.0's
vector<double> ten3s(10, 3.0 );

// Create a vector which is a copy of another
vector<double> copy( ten3s );
ASSERT( ten3s.size() == copy.size());

// replace it with myVector
copy = myVector;
ASSERT( myVector.size() == copy.size());
```

You aren't restricted to just vectors of double values. If you want an integer-valued vector, just use `vector<int>`. If you want an ordered list of strings, just use `vector<string>`. The example below shows how to initialise a vector of integers using a list of values.

```
vector<int> threeInts({ 2, 3, 4 });
```

7.2 Pass by reference and const

When you write a function that takes a vector parameter, you should write it like this:

```
double sum( const vector<double>& v ) {
    double total = 0.0;
    int n = v.size();
    for (int i=0; i<n; i++) {
        total += v[i];
    }
    return total;
}
```

Notice the strange `const` and `&` symbols that have crept in. The purpose of these additional symbols is to make it quicker to pass large objects around, as we will explain momentarily in Section 7.2.1.

I recommend that you learn the above function off by heart as a reminder of the syntax for

- Vectors

- `for` loops

Using C++ Classes 101

- Pass by reference (the &)

- The `const` keyword.

7.2.1 Pass by reference

Consider the following short program. What do you think it prints out?

```
void printNextValue( int x ) {
    x = x + 1;
    cout << "B: Value of x is "<<x<<"\n";
}

void main() {
    int x = 10;
    cout << "A: Value of x is "<<x<<"\n";
    printNextValue( x );
    cout << "C: Value of x is "<<x<<"\n";
    return 0;
}
```

The answer is that it prints out:

```
A: Value of x is 10
B: Value of x is 11
C: Value of x is 10
```

The reason is that when you call a function ordinarily, a copy is made of your data and that copy is passed to the function. In this case a *copy* of the variable x is passed to `printNextValue`. This means that the original variable x remains unchanged when the copy is incremented.

On the other hand, the very similar program:

```
void printNextValue2( int& x ) {
    x = x + 1;
    cout << "B: Value of x is "<<x<<"\n";
}

void main() {
    int x = 10;
    cout << "A: Value of x is "<<x<<"\n";
    printNextValue2( x );
    cout << "C: Value of x is "<<x<<"\n";
    return 0;
}
```

prints out something different. It prints out:

102   *C++ for Financial Mathematics*

```
A: Value of x is 10
B: Value of x is 11
C: Value of x is 11
```

The only important difference is the & symbol used in the definition of `print-NextValue2`. In this case the parameter `x` is *passed by reference*. This means that the variable x in the function main is actually modified by `printNextValue2`. The function is not passed a copy of the variable, but instead is asked to modify the variable itself.

We say that the function `printNextValue` which uses pass by value has no *side effects* since the function simply performs the computation you expect. On the other hand, `printNextValue2` has the side effect of unexpectedly changing the value of x.

As we will see in the next section, there are times when you might actually want a function to change the values of the data that you pass as parameter. When this happens, it is good practice to make sure this is obvious from the name of the function. If `printNextValue2` was instead called `increment-AndPrint` it would not be such a confusing function. Its name would make its behaviour clear. In general, programmers assume that a function won't change the parameters passed to it unless the documentation makes that very clear. As a result, just as in medicine, unwanted side effects are usually considered a bad thing.

This would seem to suggest that you should rarely use pass by reference. However, when you pass large objects around, passing data by reference can be much quicker than passing a copy by value. This is because creating a copy of a large object can take a lot of time. With pass by reference there is no need for any copying. Pass by reference works by telling the function you are calling whereabouts in memory your data can be found. Your data is read, and potentially modified, in situ.

Tip: When to use pass by reference (&)

When passing objects, pass them by reference as your default choice.

When passing the primitive data types `double`, `int`, and `bool`, pass by value is perhaps marginally quicker, so for these data types it is conventional to use pass by value.

7.2.2   The const keyword

We seem to be caught between two alternatives: Pass by reference is quick but potentially confusing; pass by value is slow but should not cause confusion. There are two possible solutions.

- Use pass by reference but take care not to actually modify the parame-

Using C++ Classes 103

ters that are passed to your function. This may seem a facetious solution to the problem, but it is a very practical approach.

- Use pass by reference and ask the compiler to *prove* that the function doesn't modify the parameters passed in. You can do this by using the `const` keyword.

When you pass a parameter as a `const` parameter, this means that the function you are calling is not allowed to change the value of the parameter. For example, if you change the code in our example so it now reads:

```
void printNextValue( const int& x ) {
    x = x + 1;
    cout << "B: Value of x is "<<x<<"\n";
}
```

you will no longer be able to compile the code. This is because the function attempts to modify the variable x contradicting the use of the `const` keyword.

You can use the `const` keyword on variables, too. We've seen it used on global variables, but you can use it on local variables. Whenever it is used, it means that the value of the variable must not change.

Tip: Use const consistently or don't use it at all

As we have discussed, using the `const` keyword means the compiler will check your functions to make sure they don't have any side effects.

This has some advantages, but it has the obvious downside that you need to type `const`! If you choose to use the `const` keyword you will quickly discover that you must use the `const` keyword consistently throughout your code in order to avoid compilation errors. It turns out that this will mean typing `const` *extremely* often.

You need to decide whether you want to use `const` in your code or not. In this book we will demonstrate how to write code using the `const` keyword and so will use `const` a lot.

If you are working on an existing C++ project, a decision will have been made on whether or not to use `const` in that project. In this case you should fall in with this decision.

If you are writing your own software project, perhaps for a student dissertation, I personally would recommend *not* using the `const` keyword. I believe that you will waste time trying to understand the compiler errors raised by `const` that could be better spent writing unit tests. However, this view is controversial and many projects do choose to use the `const` keyword.

7.2.3 Pass by reference without `const`

We have seen that pass by reference without the `const` keyword can lead to unwanted side effects. Nevertheless, there are times when you actually want a function to change its parameters.

One example is that C++ does not allow you to return multiple values. You can use pass by reference to get round this.

As an example, suppose that we want to write a function to convert the polar coordinates (r, θ) into the Cartesian coordinates (x, y). We cannot write a function that returns two doubles, so instead we write a function that doesn't return anything, but modifies two variables x and y that are passed in by reference.

```
void polarToCartesian( double r, double theta,
                       double& x, double& y ) {
    x = r*cos(theta);
    y = r*sin(theta);
}
```

Here is an example of how to use this function:

```
static void testPolarToCartesian() {
    double r = 2.0;
    double theta = PI/2;
    double x=0.0,y=0.0;
    polarToCartesian(r,theta,x,y);
    ASSERT_APPROX_EQUAL( x,0.0,0.001 );
    ASSERT_APPROX_EQUAL( y,2.0,0.001 );
}
```

As we shall see in the next chapter, user defined types provide an alternative means of returning multiple values.

7.3 Using `ofstream`

Another very useful data type is an `ofstream`. This stands for "Output File Stream". The phrase "Output Stream" is a general term for somewhere you can write data.

For example, you can write data to a file, to the screen, to a printer, over the Internet, etc. All of these give examples of output streams. In C++ you can use all these streams in much the same way.

Here is an example of how to use a `ofstream`:

```
// create an ofstream
```

Using C++ Classes

```
ofstream out;

// choose where to write
out.open("myfile.txt");

out << "The first line\n";
out << "The second line\n";
out << "The third line\n";

// always close when you are finished
out.close();
```

As you can see, it works just like `std::cout` except for the open and closing of the file. When using an `ofstream` you must call `open` to specify which file you want to write to. You must call `close` when you have finished writing.

You must call `close` because C++ performs some "buffering", which means your data might not actually be written until you call `close`. Buffering is a common performance enhancement that is used when writing data to networks and files. It is usually quicker to write 100 characters to a file in one go than it is to make a separate call for each character. As a result, when you write to a file, C++ will quietly batch together all your writes until it thinks there is enough data to be worth sending to the file. Calling `close` indicates that it is time to finish writing the data.

In addition, calling `close` indicates to other programs you are running that you are no longer trying to write to the file—you normally want to prevent two different programs trying to write to the same file.

When passing round an `ofstream`, pass a reference to an `ostream`. Read the last sentence carefully: There is no `f` in `ostream`. It stands for "output stream" and it can be used to refer to *any* kind of stream to which you might wish to write the output of your program.

```
void writeHaiku( ostream& out ) {
    out << "The wren\n";
    out << "Earns his living\n";
    out << "Noiselessly.\n";
}

void testWriteHaiku() {
    // write a Haiku to cout
    writeHaiku( cout );
    // write a Haiku to a file
    ofstream out;
    out.open("haiku.txt");
    writeHaiku( out );
    out.close();
}
```

106   *C++ for Financial Mathematics*

Notice that we pass an `ofstream` by reference but not by `const` reference. This is because we want to be able to change the stream by writing data to it.

You can pass in any kind of `ostream` to the function `writeHaiku`. We've illustrated this by writing the haiku to `cout` and to a file. But the function can be used equally well to write to a printer or across the Internet. This makes the function far more powerful. This is an example of "polymorphism", which we will discuss in detail in Chapter 10. Polymorphism is simply the Greek for "many forms": There are many different forms of output stream, but we can write functions that work with any of them.

7.4   Working with `string`

The `string` class allows you to manipulate character data conveniently. Here are some examples of the usage of a string.

```
// Create a string
string s("Some text.");

// Write it to a stream
cout << s<< "\n";
cout << "Contains "
     << s.size() <<
     " characters \n";

// Change it
s.insert( 5, "more ");
cout << s <<"\n";

// Append to it with +
s += " Yet more text.";
cout << s <<"\n";

// Test equality
ASSERT( s=="Some more text. Yet more text.");
```

We can construct a string using text in quotation marks. We can use the member function `.size` to compute its length. We can use the member function `.insert` to insert additional text at a given location. We can use `+` to add text onto the end of a string.

You should pass `string` objects using a `const` reference for efficiency.

You may recall from Section 2.4 that when you write some text in double quotation marks, this doesn't actually create an object of type `string`.

Using C++ Classes   107

It creates data of type `char*`. This means "a pointer to a memory address containing a sequence of characters". You need to tell C++ that you want to convert this to a `string`.

How does a `char*` differ from a `string`? The key difference is that a `char*` consists only of the data. A `string` combines the data and a number of helpful functions into a single package.

So, for example, the following code makes no sense in C++:

```
int a = "Quick brown fox".size();
```

You must first convert the raw data of "Quick brown fox" into a `string` object and then you can call size.

```
string s("Quick brown fox");
int a = s.size();
```

In general, we will prefer to use a `string` rather than work with the raw data directly because these functions are so helpful. Moreover, using `string` is more efficient because it stores the length of the string as well as the actual characters.

While we are discussing writing raw character strings in quotation marks, we note that you can escape characters in double quotation marks much as you can escape characters in single quotation marks. Use `\"` to mean a single quote and `\\` to mean a backslash.

You will often find that a function requires a `char*` argument rather than a `string`. In this case use the function `c_str()` to convert a `string` back into a `char *`.

This confusion between the `string` data type and the `char*` data type can be traced back to the C programming language. C never had a proper data type for strings. Programmers have historically used a pointer to a memory location containing characters to represent a string. So for historical reasons, many functions still expect you to pass in a `char*`.

`string` should be your first choice for representing character data. We'll discuss the `char*` data type more thoroughly in Chapter 11.

7.5   Building strings efficiently

Using + to build up strings by adding to the end of them is rather slow. As a result you should avoid code like this:

```
string s("");
for (int i=0; i<100; i++) {
    s+="blah ";
}
```

```
cout << s<<"\n";
```

By constantly resizing the **string** objects we waste computer effort. Instead it is better to write to a **stringstream**. A **stringstream** is an output stream that stores the characters that you write.

```
stringstream ss;
for (int i=0; i<100; i++) {
    ss<<"blah ";
}
string s1 =ss.str();
cout << s1 <<"\n";
```

Once you have built up the **string** you want, call the member function `.str()` to convert the **stringstream** to a **string**.

We won't perform much string manipulation in this book and the performance won't be a big issue for us. But using a **stringstream** is also great for testing. If you have a function that writes to an **ostream** you can test that it works by seeing if it writes to **stringstream** correctly. You can then be confident it will work when you use an **ofstream** to write to a file. You can even be confident it will work when you write to some more exotic kind of stream that represents an Internet connection.

This is another example of polymorphism and shows how polymorphism makes testing easier.

7.6 Writing a pie chart

It may surprise you to learn that we now know enough C++ to be able to produce some charts. Admittedly there is a trick: We will actually produce the *data* for a chart in C++ and then use an external program to produce the chart itself.

Given this idea, the simplest way to chart in C++ is to simply write the data to a file and then create the chart using a spreadsheet program such as Excel or OpenOffice Calc.

Here is the code needed to write a chart in a format that can be easily understood by a spreadsheet program.

```
void writeCSVChartData( ostream& out,
        const vector<double>& x,
        const vector<double>& y ) {
    ASSERT( x.size()==y.size());
    int n = x.size();
    for (int i=0; i<n; i++) {
```

Using C++ Classes   109

```
            out  <<  x[i]  <<","<<y[i]  <<"\n";
        }
    }
    void  writeCSVChart(  const  string&  filename ,
            const  vector<double>&  x,
            const  vector<double>&  y )  {
        ofstream  out;
        out.open(  filename.c_str()  );
        writeCSVChartData(  out,  x,  y  );
        out.close();
    }
```

To make this part of a library we need to declare it in the header:

```
    void  writeCSVChart(  const  std::string&  filename ,
            const  std::vector<double>&  x,
            const  std::vector<double>&  y );
```

The function declaration in the header, uses the qualification `std::` a lot.
We've had to fully qualify every type name. The reason is that you should
never write `using namespace std;` in a header file so unfortunately all these
boring `std::` prefixes are required.

Together these two functions give a very sensible solution to the problem
of writing a chart in C++. It doesn't feel very satisfying, however, because
one still has to manually turn the data into a chart using the spreadsheet
program. We can do a little better.

7.6.1   A web-based chart

It is surprisingly easy to create a web page containing a pie chart. Here
are the steps:

- Create a file called `myPieChart.html`. Open it with a text editor.

- Copy the code below into your file[1]. You can also download essentially
  equivalent code from the website:
  `https://google-developers.appspot.com/chart/`.

- Save the file.

- Open the file in a web browser.

```
<html>
  <head>
    <!--Load the AJAX API-->
```

---

[1] This code is taken from Google Charts' online documentation and altered slightly to
fit the page. The code is distributed under the Apache 2.0 license. `http://www.apache.org/`
`licenses/LICENSE-2.0`

110   C++ for Financial Mathematics

```
<script type="text/javascript"
        src="https://www.google.com/jsapi"></script>
<script type="text/javascript">

  // Load the Visualization API
  // and the piechart package.
  google.load('visualization', '1.0', {'packages':['corechart']});

  // Set a callback to run when the
  // Google Visualization API is loaded.
  google.setOnLoadCallback(drawChart);

  // Callback that creates and populates a data table,
  // instantiates the pie chart, passes in the data and
  // draws it.
  function drawChart() {

    // Create the data table.
    var data = new google.visualization.DataTable();
    data.addColumn('string', 'Topping');
    data.addColumn('number', 'Slices');
    data.addRows([
      ['Mushrooms', 4],
      ['Salami', 2],
      ['Spinach', 3]
    ]);

    // Set chart options
    var options = {'title':'Pizza Toppings',
                   'width':400,
                   'height':300};

    // Instantiate and draw our chart, passing in some options.
    var chart =
        new google.visualization.PieChart(
            document.getElementById('chart_div')
        );
    chart.draw(data, options);
  }
</script>
</head>

<body>
  <!--Div that will hold the pie chart-->
  <div id="chart_div"></div>
</body>
</html>
```

If you have done everything correctly, when you double click on the file it should show a pie chart like the one in Figure 7.1.

You may wonder how this works. In a nutshell it uses Google's web-based charting library, which makes it easy to create charts inside web documents (.html files). We haven't got time to learn HTML and Javascript, the two languages involved in creating this chart. However, this isn't a significant problem. We can simply use our intelligence to work out how to change the example code to print the pie chart of our choice.

This example code comes straight from Google's own documentation for Google Charts. If you look through their online documentation it will give you plenty of ideas about how you can change the chart to meet your needs.

We've been learning to program C++ from the bottom up. In other words we've been learning all the technical details step by step. Another route to

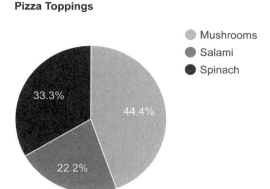

FIGURE 7.1: An example pie chart

learning a language is by immersion: you look at examples and change them. You may never learn all the strict rules of the language, but you can get a lot done with limited knowledge. So right now we're learning HTML and Javascript by immersion.

As an exercise, change the pie chart example so it produces the pie chart of your choice. Just in case it isn't obvious, the only bit of the code you need to pay attention to are the lines:

```
data.addRows([
  ['Mushrooms', 4],
  ['Salami', 2],
  ['Spinach', 3]
]);
```

The end result is that writing a pie chart won't be much harder than generating a file containing the data. There's just a bit of extra boilerplate at the top and bottom of the file that we don't need to understand.

Let us run through the entire process of creating a function to generate pie charts.

7.6.2 Create a header file

- Create a new header file.
- Call the file charts.h
- All header files should start with #pragma once

112   *C++ for Financial Mathematics*

- Include standard libraries with `#include "stdafx.h"`

- (We'll cover tests later)

```
#pragma once

#include "stdafx.h"
```

7.6.3   Write a source file

- Create a new source file

- Call the file `charts.cpp`

- All source files should `#include` the related header file

- (We'll cover tests later)

```
#include "charts.h"
```

7.6.4   Enable testing in your files

In `charts.h`

```
void testCharts();
```

In `main.cpp`

```
int main() {
    testMatlib();
    testGeometry();
    testCharts();
    testUsageExamples();
}
```

In `charts.cpp`

```
void testCharts() {
}
```

7.6.5   Write functions to generate the boiler plate

- We pass an `ostream&` reference to the function

- We use `\"` to escape quotes in quotes.

Using C++ Classes
113

- The spacing in HTML files isn't very important, so this function doesn't reproduce the spacing of Google's example pie chart precisely. This was done so we could fit the code on the page.

```cpp
static void writeTopBoilerPlateOfPieChart( ostream& out ) {
    out<<"<html>\n";
    out<<"<head>\n";
    out<<"<!--Load the AJAX API-->\n";
    out<<"<script type=\"text/javascript\"";
    out<<"src=\"https://www.google.com/jsapi\">";
    out<<"</script>\n";
    out<<"<script type=\"text/javascript\">\n";
    out<<"google.load('visualization', '1.0',";
    out<<" {'packages':['corechart']});\n";
    out<<"google.setOnLoadCallback(drawChart);\n";
    out<<"function drawChart() {\n";
    out<<"var data=new google.visualization.DataTable();";
    out<<"\n";
    out<<"data.addColumn('string', 'Label');\n";
    out<<"data.addColumn('number', 'Value');\n";
}
```

Writing a function for the bottom boiler plate code is just as easy.

```cpp
static void writeBottomBoilerPlateOfPieChart(
        ostream & out ) {
    out<<"var options = {'title':'A Pie Chart',\n";
    out<<"'width':400,\n";
    out<<"'height':300\n";
    out<<"};\n";
    out<<"var chart = new google.visualization.PieChart(";
    out<<"document.getElementById('chart_div'));\n";
    out<<"chart.draw(data, options);\n";
    out<<"}\n";
    out<<"</script>\n";
    out<<"</head>\n";
    out<<"<body>\n";
    out<<"<div id='chart_div'>\n";
    out<<"</body>\n";
    out<<"</html>";
}
```

7.6.6 Write a simple version of the chart data

The hardest bit of code will be writing out the pie chart data. For the time being we "cheat" and just write a simplified function that will print out some fixed text that we hope will work.

114　　　　　　　　*C++ for Financial Mathematics*

This isn't really cheating, just a sensible practice. We work in small pieces. Once you've solved one simple problem, move on to the next simple problem.

```
static void writeFixedPieChartData( ostream& out) {
    out<<"data.addRows([\n";
    out<<"['Bananas', 100],\n";
    out<<"['Apples', 200],\n";
    out<<"['Kumquats', 150]\n";
    out<<"]);\n";
}
```

7.6.7　Write a test of what we've done so far

```
static void testFixedPieChart() {
    ofstream out;
    out.open("FixedPieChart.html");
    writeTopBoilerPlateOfPieChart(out);
    writeFixedPieChartData( out );
    writeBottomBoilerPlateOfPieChart( out );
    out.close();
}
```

```
void testCharts() {
    TEST( testFixedPieChart );
}
```

We've written enough code to test, so let's run it. When you run this test using the standard development environment set-up, it will create the file in the same folder as `main.cpp`.

7.6.8　Write the interesting code

Let us be clear on what the interesting bit of code must do. Given a string of labels it should produce output that looks like this:

```
data.addRows([
['Bananas', 100],
['Apples', 200],
['Kumquats', 150]
]);
```

The last line is special—there is no comma. Other than that it all looks straightforward.

```
static void writeDataOfPieChart( ostream& out,
```

Using C++ Classes

```cpp
                        const vector<string>& labels,
                        const vector<double>& values) {
    out << "data.addRows([\n";
    int nLabels = labels.size();
    for (int i=0; i<nLabels; i++) {
        string label = labels[i];
        double value = values[i];
        out<<"['"<<label<<"', "<<value<<"]";
        if (i!=nLabels-1) {
            out<<",";
        }
        out<<"\n";
    }
    out<<"]);\n";
}
```

There is one subtle problem. For simplicity we have assumed that the labels don't contain quotation marks or other special characters. We leave it as an exercise to fix this issue.

Danger!

It is important to fix issues with special characters in production software. Hackers love bugs based around quotation marks. For example, when using a web application, they might try choosing the peculiar user name

```
Dave'); DELETE FROM Users; --
```

Here the hacker is guessing that you are using an SQL database and they are hoping to trick you into executing the following SQL code on your database[2]:

```
INSERT INTO Users VALUES ('Dave'); DELETE FROM Users; --')
```

This code would first create a new User called "Dave", but then would delete the entire database table Users! The issue is that the ' character has a special meaning in SQL just as it does in C++ and so it needs to be escaped when it is used as part of a string.

7.6.9   Testing the interesting code

To test the function `writeDataOfPieChart` we write a test that finds the text output by the function `writeDataOfPieChart` and then compares it to the expected value. To do this we first set up some dummy data for our pie chart. We then call `writeDataOfPieChart` and store the output using a `stringstream`.

116   *C++ for Financial Mathematics*

```cpp
static void testPieChartData() {
    // this test automates the checking
    stringstream out;
    vector<string> labels(3);
    vector<double> vals(3);
    for (int i=0; i<3; i++) {
        stringstream ss;
        ss<<"A Label "<<i;
        labels[i] =ss.str();
        INFO( labels[i] );
        vals[i]=(double)i;
    }
    writeDataOfPieChart( out,
                labels,
                vals );
    string asString = out.str();
```

We can now compare the output of `writeDataOfPieChart` to the output we are expecting.

```cpp
    stringstream expected;
    expected<<"data.addRows([\n";
    expected<<"['A Label 0', 0],\n";
    expected<<"['A Label 1', 1],\n";
    expected<<"['A Label 2', 2]\n";
    expected<<"]);\n";
    string expectedStr = expected.str();
    ASSERT( asString==expectedStr );
}
```

This gives a test that `writeDataOfPieChart` behaves in the way that we want.

- Since `fstream` and `stringstream` are both types of `ostream`, the interesting code was easy to test.

- You might have thought it would be impossible to write a useful test for code that generates a chart. Many people would argue that you need a human to look at the chart to test the code. However, we have just shown that it is perfectly possible to write meaningful tests of almost any code. If you can't test it, you've designed your code incorrectly or don't understand the problem properly.

7.6.10   Wrap it all up into a single function

The function definition below makes it easy to create a pie chart with given labels and values from C++. The pie chart will be saved in the given file.

```
void pieChart( const  string& file,
               const  vector<string>& labels,
               const  vector<double>& values ) {
   ofstream out;
   out.open(file.c_str());
   writeTopBoilerPlateOfPieChart(out);
   writeDataOfPieChart( out, labels, values );
   writeBottomBoilerPlateOfPieChart( out );
   out.close();
}
```

Hopefully you will agree that this function is very simple. We have used functions to divide our code into manageable pieces which together produce an interesting result.

If we want to make our function available in other files, we will need to put a declaration into the header files as follows:

```
void pieChart( std::string& file,
               std::vector<std::string>& labels,
               std::vector<double>& values );
```

This is copied from the code in the `.cpp` file except we've had to put in lots of `std::` statements since you should never write `using namespace std;` in a header file.

7.7   The architecture of the World Wide Web

The Internet and the World Wide Web are slightly different things. The Internet includes everything on the Internet including Skype, email and Netflix. The World Wide Web is what you can see in a web browser.

So, the World Wide Web is just one particularly popular application that runs on the Internet.

Our "trick" of writing a chart as a text-based file relies on the architecture of the World Wide Web. The World Wide Web is based on sending text files around. Servers receive text data (from users filling in forms and typing URLs). Servers send out text data (HTML files like our chart).

It's easy to write the server that generates a web page without even understanding HTML. We've just done this ourselves. It is also easy to test your server code without even using a web browser.

This is a "design pattern" that you should copy. Use human-readable data to communicate between applications. Write small simple applications that communicate through text files. This will allow you to use C++ for what C++

118   *C++ for Financial Mathematics*

is good at and other languages where they work better (e.g., user interfaces and prototyping).

This is the first significant example of *software architecture* in this book. Developing software may require you to do any of the following:

- write code in a given language;

- design functions and data types in that language to produce convenient and useful libraries;

- write systems using multiple computers and languages to create a useful end product.

These tasks are termed coding, software design, and software architecture, respectively. Tim Berners-Lee was acting as a software architect when he devised the World Wide Web. As the World Wide Web demonstrates, a good architecture can make all the difference to the success of your application.

An introductory book like this on programming in C++ inevitably focuses on coding and on software design. However, it is worth keeping in mind that a good software developer will be willing to use multiple interacting programs and multiple languages to achieve their goals.

Finally, you may wonder how to produce charts that you can include in publications from your C++ code. A tool such as GNU plot can be used to produce very high-quality charts. The essential idea is similar: use C++ to write a text file GNU plot understands; run GNU plot to turn this text file into a pdf file ready for publication.

Exercises

7.7.1. Write a function `solveQuadratic` which takes parameters a, b, and c and which uses the quadratic formula to find both roots of $ax^2 + bx + c = 0$. It should return the number of roots and use pass by reference to return the values of the roots. Write an alternative version of this function which uses a `vector` to return the answer.

7.7.2. Write a function `mean` that computes the mean of a vector of doubles. Write a test for this function.

7.7.3. Write a function `standardDeviation` that computes the standard deviation of a vector of doubles. Make it take two parameters, the vector itself and a boolean indicating whether to compute the population standard deviation or sample standard deviation. Make the default value be to compute the sample standard deviation. Write a test for this function. Why would `std` be a bad choice of name for this function?

Using C++ Classes

119

7.7.4. Write a function `min` and a function `max` which each take a vector of doubles and return the max and min, respectively. Write tests for these functions. Have you written a helper function to create a test vector yet?

7.7.5. Write a function `randuniform(int n)` which returns a vector of uniformly distributed random numbers in the range $(0, 1)$. Use the function `rand()`, which you will need to #include from `cstdlib` (look up the documentation here: `http://www.cplusplus.com/reference/cstdlib/rand/`) to complete this exercise. You will also need to use the constant `RAND_MAX` that is also defined in `cstdlib`. Write a test for this function.

Note that in this book we won't try to tell you every function you might want to use, so it is a good idea to start becoming familiar with how to use the online documentation.

7.7.6. Write a function `randn(int n)` which returns a vector of normally distributed random numbers with mean 0 and standard deviation 1. Write a test for this function.

7.7.7. An alternative way to generate normally distributed random numbers is to use the Box–Muller algorithm. To do this one first generates two uniformly distributed random variables u_1 and u_2 in the interval $(0, 1)$. Define:

$$n_1 = \sqrt{-2 \log(u_1)} \cos(2\pi u_2)$$

$$n_2 = \sqrt{-2 \log(u_2)} \sin(2\pi u_2)$$

n_1 and n_2 will be independent normally distributed random variables with mean 0 and standard deviation 1. Write a function to generate N normally distributed random numbers using the Box–Muller method and write a test for your answer.

7.7.8. Use the Google Line Chart example to find out how to create a web page with a line chart in it. Use this to write a function `plot` which takes a string, a vector of x coordinates, and a vector of y coordinates, and generates a line chart containing a graph of y against x. Test your code by plotting a graph of $y = x^2$.

Plot a graph of the price of a call option against the current stock price using Equation (A.6). Assume the volatility is 0.2, the strike price is 100, the time to maturity is 1.0, and the risk-free interest rate is 0.05.

7.7.9. You can use the standard library function `sort` to sort a vector of doubles. If `v` is a vector, you can sort it using the following code:

```
sort( v.begin(), v.end() );
```

You will also need to #include `<algorithm>` and use the namespace `std`. The code above actually changes the vector `v`, so you may prefer to create a sorted copy as follows:

```
vector<double> copy = v;
sort( copy.begin(), copy.end() );
```

Use this to write a function `prctile` that takes as input a vector of doubles v and a percentile p and outputs the p-th percentile.

Note that there is no agreement on the precise definition of percentile. You will need to choose a precise definition to answer the question. The definition used in the model answer given for this exercise is illustrated in Figure 7.2 for the sample vector $(1\,3\,4\,7)$. In this definition of a percentile, if v contains n

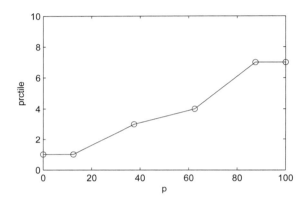

FIGURE 7.2: Percentiles of the vector $(1\,3\,4\,7)$

points, then the sorted values of v determine the percentiles at p values equal to:
$$100 \times \frac{1}{2n},\ 100 \times \frac{3}{2n},\ 100 \times \frac{5}{2n},\ \ldots,\ 100 \times \frac{2n-1}{2n}.$$
The 0-th percentile is always equal to the smallest value in the sample. The 100-th percentile is the largest value in the sample. All other percentiles are determined by linear interpolation.

7.7.10. (**More challenging**) You will find that the Google Line Chart can be used to plot curves other than just the graph of functions. This is because the vector of x coordinates need not be in ascending order and can even contain duplicates. Use this idea to write a function `hist` that takes a vector of numeric values and a number, n, of buckets and plots a histogram containing n bars that shows the frequency of the different values. Your code should simply create the data needed to plot a line chart and then use the answer to the previous question. An alternative approach would be to use the Histogram functionality provided by Google charts. What would be the problem with this second approach for very large samples?

Using C++ Classes   121

7.7.11. **(More challenging)** Change the code so that it can cope with special characters such as ' inside the labels. You will want to find out how to look at strings a character at a time in the online documentation for `string`, which you can find on `cplusplus.com`.

7.7.12. In the above questions, have you passed vectors and strings by reference? Have you used the `const` keyword where possible? Have you used the `static` keyword where possible?

7.8   Summary

This chapter marks something of a transformation of our C++ programming skills. By putting everything we've learned together, we can write something very sophisticated.

The specific skills we learned were:

- Passing by reference and passing by value.

- Using the `const` keyword if we want the compiler's help on avoiding side effects.

- Using `vector<double>` to represent a mathematical vector. Pass them as the type `const vector<double>&`.

- Using `string` to represent strings. Pass them as `const string&`.

- Using `stringstream` to build complex strings.

- Using `fstream` to write to files.

- Using `ostream` to write to any kind of stream interchangeably.

- Using the architectural design pattern of communicating between different programs using text files.

Chapter 8

User-Defined Types

In the previous chapter we learned how to use:

- `vector<double>`,

- `string`,

- `ofstream`,

- `stringstream`.

Using these new types has made us much more productive C++ programmers. The next step is to learn to write our own types, which we hope will similarly boost our productivity.

As you will learn, the main job of a programmer in an object-oriented language is writing their own types.

Throughout this chapter we will assume that you have opened FMLib8 in your development environment so you can view the code and experiment with it.

8.1 Terminology

We already know the following.

- Every variable in C++ has a type (for example, `double`).

- You can have lots of variables of a given type.

We are about to start learning a programming style called "object-oriented programming" in which it is conventional to use slightly different vocabulary for the same ideas.

- Every *object* in C++ has a *class*.

- You can have lots of *instances* of the same class.

Example 1:

123

124   *C++ for Financial Mathematics*

```
string s("To be, or not to be?");
```

In the code above, s is an *object* of type string. It is an *instance* of the *class* string.

The term *object* in object-oriented programming refers to a data type which consists of a bundle of data together with helpful functions to work with that data.

So a vector<double> is an object because it contains the data of a vector and helpful functions like size(). Similarly a string is an object.

On the other hand, double, int and the raw data type char* are not object-oriented types because they consist purely of data with no special helper functions.

The type of data and functions supported by an object depend only on its *class*. All instances of the same class have the same functions and the same types of data.

Example 2: An instance of the class string consists of:

- data consisting of a sequence of characters;

- functions such as size, insert, erase for working with these data.

The *class* string is where all these functions and data types are defined.

Finally, the functions associated with an object are often called *methods*. There isn't meant to be any meaningful difference between the words function and method. Some authors prefer the word method as the word function can scare non-mathematicians. We will use both terms.

8.2   Writing your own class

8.2.1   Writing the declaration

To show how to write your own class, let us create a data type which represents a point in two-dimensional space.

```
class CartesianPoint {
public:
    double x;
    double y;
};
```

User-Defined Types 125

The code above is called a class declaration. You can choose whether to put a class declaration in a `.h` file or a `.cpp` file. If you want it to be part of your library, you put it in the header file. If you are using it to implement your library but don't want anyone else to use it, you put it in the `.cpp` file.

In this case we want the class to be available to users of our library, so we've declared it in the `.h` file `geometry.h`.

In the example above:

- The name of our class is `CartesianPoint`.

- A Cartesian point contains two `double` values called x and y. These are called *member variables* of the class.

- Users of the library are allowed to use the values of x and y in their own code, so these are marked as `public`.

When writing classes one uses the following general syntax for the class declaration:

```
class CLASS_NAME {
public:
    DATA AND FUNCTION DECLARATIONS
private:
    DATA AND FUNCTION DECLARATIONS
};
```

The data and functions you declare in the section marked `public:` will be available to users of your library. The data and functions marked as `private:` are not available to users of your library. This allows you to hide information about how your class works. This will make your library easier to use and maintain.

Some points to notice are:

- You can define classes in header files or `cpp` files. Use header files if you want users of the library to use your class.

- You must remember the semi-colon at the end. You may get very confusing compiler errors if you omit it.

- Like variable names, class names should contain no spaces or strange characters and should start with a letter. It is a good idea to use a naming convention so it is easy to distinguish class names and variable names. In this book we'll use the naming convention of writing class names using camel case starting with a capital letter. For example, we have the class name `CartesianPoint`.

8.2.2 Using a class

Let us see how someone will use the `CartesianPoint` class. Our example code contains a test that shows how the class can be used.

```
1     CartesianPoint p;
2     p.x = 100;
3     p.y = 150;
4     cout << "Coordinates (";
5     cout << p.x ;
6     cout << ", ";
7     cout << p.y ;
8     cout << ")\n";
9
10    p.x *= 2;
11    p.y *= 2;
12    cout << "Rescaled cordinates (";
13    cout << p.x ;
14    cout << ", ";
15    cout << p.y ;
16    cout << ")\n";
```

This creates a point and sets the coordinates to $(100, 150)$. It then prints out the coordinates. Next it doubles all the coordinate values and prints them out a second time.

To understand this in more detail, at line 1 we create an instance of the class `CartesianPoint`. We do this simply by declaring a variable whose type is `CartesianPoint`.

```
CartesianPoint p;
```

We have now created a variable called `p` that represents a point in space. We then set the member variables representing the x coordinate and the y coordinate. We choose the values so that p represents the point $(100, 150)$.

```
p.x = 100;
p.y = 150;
```

When you want to access the data inside the class you use a dot . followed by the name of the member variable.

You can only access the variables because they have been marked as `public:`. Try removing the word `public:` and see what compiler error you get. This will help you understand the compiler error in future should you ever happen to forget to mark your data as public.

8.2.3 Passing objects between functions

Let us write a new class called `PolarPoint` which represents a point in polar coordinates. It will therefore have member variables corresponding to polar coordinates (r, θ).

```
class PolarPoint {
public:
    double r;
    double theta;
};
```

We can use this class to illustrate how one passes objects to and from functions. Let us write a new function `polarToCartesian` which takes a point in polar coordinates and returns the corresponding `CartesianPoint`.

```
CartesianPoint polarToCartesian(const PolarPoint& p){
    CartesianPoint c;
    c.x = p.r*cos( p.theta );
    c.y = p.r*sin( p.theta );
    return c;
}
```

As we discussed in Section 7.2.1, you should use pass by reference when passing objects to functions. This explains the use of the `const` keyword and the `&` symbol. Normally any custom data type should be passed by reference. Pass by value is normally reserved just for `double`, `int`, and `bool`.

8.2.4 How have classes helped?

It is worth pausing to see how using classes has improved our code. Let us examine the code needed to test our new `polarToCartesian` function:

```
static void testPolarToCartesian() {
    PolarPoint p;
    p.r = 2.0;
    p.theta = PI/2;
    CartesianPoint c = polarToCartesian( p );
    ASSERT_APPROX_EQUAL( c.x,0.0,0.001 );
    ASSERT_APPROX_EQUAL( c.y,2.0,0.001 );
}
```

and compare it with our previous test code:

```
static void testPolarToCartesian() {
    double r = 2.0;
    double theta = PI/2;
    double x=0.0,y=0.0;
```

128   *C++ for Financial Mathematics*

```
    polarToCartesian(r,theta,x,y);
    ASSERT_APPROX_EQUAL( x,0.0,0.001 );
    ASSERT_APPROX_EQUAL( y,2.0,0.001 );
}
```

The first advantage that using classes has given us is that they have established a clear convention that x-coordinates are called x and y-coordinates are called y. This will make code written using CartesianPoint clearer than code that uses x some of the time, X some of the time, and xcoord the rest of the time.

The second advantage that using classes has given us is that our code no longer depends so heavily on the ordering of parameters. One needs to be very careful in the original code that you pass every parameter in the correct order, and the compiler will provide no help. With classes it becomes impossible to get the order wrong. This becomes particularly important when, for example, passing 5 parameters to a function to price a call option. It is then all too easy to pass the parameters in the wrong order.

The third advantage is that we are no longer using pass by reference to return multiple values. Most people find returning a single object conceptually simpler than modifying parameters by reference. This means that the version of the code using classes is easier for most people to understand.

So using classes gives us consistency, readability, and reliability. The advantages may seem marginal at the moment, but, as our examples become more and more complex, the advantages will become increasingly apparent.

There is a trade-off, however. We've had to write the code to represent the classes. We should ask whether this trade-off is worthwhile. The philosophy of library design says that our aim should always be to make life easy for the users of the library even at the expense of more work for the author of the library. This is because a library that is any good will be used a lot and written only once. So it is worth putting in some extra work to make a library as good as possible. According to this philosophy, therefore, the trade-off is worthwhile.

In this example, the advantages of object orientation are quite small. As we learn more of the object-oriented features of C++, the advantages of this programming style will increase.

8.3   Adding functions to classes

The value of object orientation becomes more apparent when we add *member functions* to our classes. Let us illustrate this with a `Circle` class. Our circle class will have only one piece of data associated with it, the radius of the circle. But it will also have various functions for working with that data.

User-Defined Types 129

We will write a function `area()` to compute the circle's area and a function `circumference()` to compute its circumference.

```
class Circle {
public:
    double radius;
    double area();
    double circumference();
};
```

We put the declarations for the functions `area` and `circumference` inside the declaration of the class `Circle`. They're marked as `public` because we want them to be available to users of our library. The declarations look like ordinary function declarations. Notice that they don't take any parameters at all.

Before we attempt to implement the functions, let us be test-driven developers and write tests for them. This will allow us to illustrate how you call the functions on an object.

```
static void testAreaOfCircle() {
    Circle c;
    c.radius = 4;
    ASSERT_APPROX_EQUAL( c.area(), 16*PI, 0.01 );
}

static void testCircumferenceOfCircle() {
    Circle c;
    c.radius = 2;
    ASSERT_APPROX_EQUAL(c.circumference(),4*PI,0.01);
}
```

As the examples above show, you use a dot to call a function on an object. For example, `c.area()` computes the area of the circle c.

We are calling the area function on a specific circle c, so the function `area` should return the area of that specific circle c. This is why you don't need to pass the radius to the member function `area`.

Now that we know how the functions will be used, let us see how to provide definitions for these functions. As usual for C++ functions, we write the definitions in a `.cpp` file.

```
double Circle::area() {
    return PI*radius*radius;
}
```

```
double Circle::circumference() {
    return 2*PI*radius;
}
```

130   *C++ for Financial Mathematics*

There are several things to observe about these function definitions.

(i) When you define a member function of a class, you must always specify the name of the class in the definition. The syntax is: `CLASS_NAME::-FUNCTION_NAME`. If you forget to do this you'll get linker errors saying you've forgotten to define the function. The reason why you must specify the name is that many classes may have functions of the same name. For example, we might have a class `Square` which also has an `area` function. This is why you need to say which class's method you are actually implementing.

(ii) The `Circle` class's member functions are able to access the member variable `radius` and the global variable `PI`.

8.3.1   Using const on member functions

As we mentioned when the `const` keyword was introduced in Section 7.2, if you are going to use `const` at all, you must use it consistently throughout your code. If you have decided to use `const`, then if a member function doesn't change the object, you should mark it as `const`. You do this by writing `const` at the end of the declaration and definition.

```cpp
class Circle {
public:
    double radius;
    double area() const;
    double circumference() const;
};

/**
 *   Computes the area of a circle
 */
double Circle::area() const {
    return PI*radius*radius;
}
```

In this example we note that measuring the area of a circle doesn't change the circle in any way. This is why we have added `const` to the declaration of the `area` function.

We have then added it to the definition as well because the use of `const` and `&` must exactly match between a functions declaration and definition. If they don't match exactly, you will get linker errors. It is worth experimenting with seeing what errors actually occur if you don't have matching `const` statements.

When you pass a reference to `const Circle` as a parameter to a function, the compiler will insist that the function does not modify that circle. This means that you won't be able to call any of the functions on the circle except

User-Defined Types 131

those that are marked as `const`. This is why, if you intend to use `const` at all, you must use it on every function that does not change the object.

8.4   A financial example

To see the benefit we gain from the introduction of classes, let us work through a more sophisticated example involving financial mathematics. First we define a class called `BlackScholesModel`.

```
class BlackScholesModel {
public:
    double stockPrice;
    double volatility;
    double riskFreeRate;
    double date;
};
```

The choice of member variables in this class is carefully considered. This class only contains the variables associated with the model and not variables associated with the financial contract. We will specify different option contracts in different classes shortly.

We will add functionality to this class in later chapters. For example, we will add functions to do things like simulate option prices.

Notice that we're specifying the date as a `double`. We'll measure dates in years since 0 A.D. So January the first 2014 would be represented as 2014.0. You wouldn't do this in real code, but we don't want to waste energy thinking about leap years and how many days September hath. When you come to price real option contracts, you will want to be a little more careful. You will then want to use a library containing a proper date class.

We now define a separate class for the option contract. We'll create a class `CallOption` which contains a strike and a maturity but does not contain any details about the current market data. The idea behind this design is that we can use the same `BlackScholesModel` to price either a `CallOption` or a `PutOption`. Note that the `CallOption` does not change as the market data changes. This reflects reality: a call option *contract* is unchanged by the market even though its price may vary.

```
class CallOption {
public:
    double strike;
    double maturity;

    double payoff( double stockAtMaturity ) const;
```

```
        double price( const BlackScholesModel& bsm )
            const;
};
```

The `CallOption` has two functions. One function, `payoff`, computes the payoff at maturity. The second, `price`, computes the price given some hypothetical market data in the form of a `BlackScholesModel`.

We've been careful with the `const` statements here. Pricing a `CallOption` does not change it, nor does it change the model used to price the option.

Let us now examine the implementation code for the `payoff` function:

```
double CallOption::payoff(
        double stockAtMaturity ) const {
    if (stockAtMaturity>strike) {
        return stockAtMaturity-strike;
    } else {
        return 0.0;
    }
}
```

Notice that all the `const` keywords exactly match the declaration. If there were any mismatch the code would not link correctly. Apart from the `const` keyword and the `CallOption::`, this code is quite straightforward. The definition of the function simply follows from the definition of a CallOption contract.

The next function is more sophisticated in that it prices the `CallOption` using the `BlackScholes` formula.

```
double CallOption::price(
        const BlackScholesModel& bsm ) const {
    double S = bsm.stockPrice;
    double K = strike;
    double sigma = bsm.volatility;
    double r = bsm.riskFreeRate;
    double T = maturity - bsm.date;

    double numerator =
        log( S/K ) + ( r + sigma*sigma*0.5)*T;
    double denominator = sigma * sqrt(T );
    double d1 = numerator/denominator;
    double d2 = d1 - denominator;
    return S*normcdf(d1) - exp(-r*T)*K*normcdf(d2);
}
```

Again note that we specify this is the `price` function for a `CallOption` using `CallOption::`. We pass the `BlackScholesModel` by `const` reference as specified in the declaration of the function for the usual efficiency reasons.

Similarly we specify that the function does not change the CallOption itself using the **const** keyword exactly as in the declaration.

Last but not least, we want a unit test. As well as checking the code, this provides a helpful example of how to use our classes.

```
static void testCallOptionPrice() {
    CallOption callOption;
    callOption.strike = 105.0;
    callOption.maturity = 2.0;

    BlackScholesModel bsm;
    bsm.date = 1.0;
    bsm.volatility = 0.1;
    bsm.riskFreeRate = 0.05;
    bsm.stockPrice = 100.0;

    double price = callOption.price( bsm );
    ASSERT_APPROX_EQUAL( price, 4.046, 0.01);
}
```

8.4.1 What have we gained?

The benefits are similar to those from introducing a `CartesianPoint` and a `PolarPoint` class, but they are now rather more striking.

- **Easier programming**. If we have a single function `blackScholesPrice` that takes 5 double parameters, it is almost impossible to remember what the correct order for the parameters is.

- **Easier debugging**. If you have a function that takes 5 double parameters, its almost impossible to spot if someone has accidentally put the parameters in the wrong order.

- **Consistency**. If we use the same `BlackScholesModel` class to price put options, Asian options, knock-out options, etc., we'll have a library that is much easier to use.

In later chapters, we will apply more advanced object-oriented techniques which will bring more striking benefits.

8.5 Recommendations on writing classes

You have many choices available to you when you write a class about such matters as the name of the class and which file to define it in. It is good to have some conventions. In this book we choose to follow the following guidelines.

1) Whenever possible, don't put class declarations in header files, put them in `cpp` files. This follows from the general principle that you should try to hide information if possible.

2) If you decide to put a class in a header file, define only one class in each header file.

3) Name that class the same as the header file.

4) Give classes names that are nouns: for example `CartesianPoint` or `Black-ScholesModel`.

5) Use upper case for the first letter of a class name.

The files `BlackScholesModel.h` and `CallOption.h` demonstrate these conventions.

The file `geometry.h` breaks these conventions but only because we don't want to clutter the project with too many files that have nothing to do with financial mathematics. If we were really writing a geometry library, we would put all the point classes in their own header files for readability.

A second issue to consider is the `const` keyword. In this chapter we have tried to use the `const` keyword consistently. However, the `const` keyword is an optional feature in C++. You can choose to use it if you want the compiler to guarantee that you never accidentally change an object when you didn't mean to, but your code will work if you never use `const` at all.

This means that you can make a software architecture decision for your project. Will you use `const` or won't you? This is a decision for the entire project because it turns out that if you want to use the `const` keyword at all, you really need to use it everywhere. This is because if someone has passed you a `const` reference, you can only pass it to functions that themselves are declared as taking a `const` reference.

In practice, this means you won't get to choose whether to use `const` or not. You will need to ask your boss whether she wants you to use it or not.

Even if you don't like the `const` keyword yourself, if you are writing a library, you'll need to use the `const` keyword if you your library users want to use the `const` keyword.

In this book we will use `const` so that you can understand the issues raised, but you may prefer to avoid using it in your own code.

8.6 Encapsulation

Let us now try to understand the `public` keyword in more detail. To do this let's write an object-oriented version of our pie chart code. We begin with a class declaration called `PieChart`.

```
class PieChart {
public:
    void setTitle( const std::string& title );
    void addEntry( const std::string& label,
                        double value );
    void writeAsHTML( std::ostream& out ) const;
    void writeAsHTML(const std::string& file ) const;
private:
    std::string title;
    std::vector<std::string> labels;
    std::vector<double> values;
};
```

Following our conventions, you can find this in `PieChart.h`.

We have now marked the data variables as `private`. This means only member functions of `PieChart` can see those variables.

The only member function that modifies the title is `setTitle`. The only function that modifies the list of labels and the list of data points is `addEntry`. Here are their definitions:

```
void PieChart::setTitle( const std::string& t ) {
    title = t;
}

void PieChart::addEntry( const string& label,
                                double value ) {
    labels.push_back( label );
    values.push_back( value );
}
```

The interesting point is that it is now impossible for `labels` and `values` ever to contain a different number of elements. This is because the only code that ever modifies them is `addEntry` and that function maintains the guarantee that the two vectors always have the same length. We know that there can't be any other code elsewhere in the project that breaks this guarantee because the variables are `private`. Because we have used the `private` keyword, we only have to check that member functions are well behaved to be certain that `labels` and `values` have the same number of elements.

It is considered good programming style to make *all* member variables private on your classes. Doing this allows you to guarantee that your object always remains in a consistent state. In addition it allows you to change your mind in the future about the implementation details (i.e., how data is stored) without users of your class being affected in anyway.

In summary, the `private` keyword allows you to perform more sophisticated information hiding than just choosing what to put in the header file.

Not putting a class in a header file gives even more information hiding than making things `private`, so that should always be your first choice. However, if you have to make the class available to users of your library, you should try to make all its data, and as many of its member functions as possible, `private`.

It was only for the sake of simplicity that we have given the `BlackScholesModel` and the `CallOption` public data members. We wanted to illustrate the use of `public` data before explaining about `private` data. In practice the library could be improved by modifying these classes.

The idea of hiding data using the `private` keyword is often called *encapsulation*. The word encapsulation really refers to two things:

(i) the bundling together of related items into a single object;

(ii) preventing direct messing with the internal data of an object.

Encapsulation is common in real-world design as well as in object-oriented software design. For example, consider the design of a car.

In a car, all the lighting controls are put on the dashboard and are clearly separated from the controls for the windows and the seats. This grouping of functionality in a car's controls corresponds to the grouping of functionality into different classes in software design.

In addition, you can control a car through standard functions (turn left, turn right, etc.) but the internal workings of a car are hidden from the user completely (they're kept under the bonnet). This corresponds to the software design principle of having a small number of public functions that are available to your users, but keeping most of the details private.

Cars are much easier to use because of the use of encapsulation. It is a good thing that you can drive a car without being a trained mechanic. Just as these design principles make cars easier to use, so the same principles make objects in object-oriented programs easy to use.

To illustrate this further, consider the question of how the `vector<double>` class stores its data given that it can't be using a `vector<double>` itself. You don't know how this is done, and quite reasonably you probably don't care that much. You don't need to know and your life has been made easier by not having to think about this.

The designer of the C++ libraries has tried to make life easier for you by making the internal data of a `vector` `private`. You should help users of your library in the same way.

8.6.1 Implementing PieChart

You can find the full implementation for the PieChart class in `PieChart.cpp`. It is simply a re-organisation of the non-object-oriented code we wrote in Section 7.6. You can call our new `PieChart` class an object-oriented *wrapper* around our existing code that makes it easier to use.

8.6.2 Using PieChart

A central belief in object-oriented programming is that human beings find the code below easy to understand:

```
static void testPieChartClass() {
    // just checks that the class compiles etc.
    PieChart pieChart;
    pieChart.addEntry("Mushrooms",200);
    pieChart.addEntry("Salami",100);
    pieChart.addEntry("Spinach",150);
    pieChart.setTitle("Pizza Toppings");

    pieChart.writeAsHTML( "PizzaPie.html" );
}
```

This really summarises one of the key points of object orientation. Object-oriented software should make life easier for users of our libraries.

Exercises

8.6.1. Write a class `PutOption` that behaves in a similar fashion to a `CallOption`. Make sure you write appropriate tests and follow the conventions on naming files.

8.6.2. Write a class `LineChart` which allows you to create a line plot. Add a feature to set the title of the plot.

8.6.3. Add a function `distanceTo` to the `CartesianPoint` class. It should take a single parameter which is another `CartesianPoint` and it should compute the distance between the two points. Here is an appropriate test function:

```
static void testDistanceTo() {
    CartesianPoint p1;
    p1.x = 1;
    p1.y = 1;
    CartesianPoint p2;
```

138   *C++ for Financial Mathematics*

```
    p2.x = 4;
    p2.y = 5;
    double d = p1.distanceTo( p2 );
    ASSERT_APPROX_EQUAL( d, 5.0, 0.0001);
}
```

Have you used the `const` keyword twice in the declaration and twice in the definition?

8.6.4. Write an ordinary function `perimeter` which takes `const` references to three CartesianPoints and computes the perimeter of the resulting triangle. The `perimeter` function should call `distanceTo`. Test your `perimeter` function with a 3, 4, 5 triangle.

What happens if you remove either of the `const` keywords from the definition of `distanceTo` but leave them in the definition of `perimeter`? Explain why it is that "if you use the `const` keyword at all, you have to use it everywhere".

8.7   Constructors

If you create a variable of type `double` in C++ but do not set its value yourself, C++ does not provide any guarantees what value will be assigned. In practice, the computer will just grab some free memory and use whatever values happen to be there.

For example, the program below is valid C++, but it isn't possible to say in advance what it will print out.

```
int main() {
    double d;
    cout << "What is the value of d?\n";
    cout << d;
    return 0;
}
```

This hasn't been an issue because we've been paying heed to warning messages and this will certainly cause a compiler warning even though it is technically valid C++ code.

The following code will also result in a compiler warning for similar reasons.

```
class Point {
public:
    double x;
    double y;
};
```

User-Defined Types   139

```
int main() {
    Point p;
    cout << "What is the value of x?\n";
    cout << p.x;
    return 0;
}
```

C++ would be a little easier to use if all `double` values and all `int` values defaulted to 0. We can't fix C++ to make things easier, but we can make our own classes easier to use by giving them *constructors*. A constructor performs the initialisation of an object to leave it in a sensible state.

We've already used constructors. For example, to construct a vector of length 100 initialised with zeros, we know we should write:

```
vector<double> v(100,0.0);
```

Similarly, to construct a string with characters `Some text`, we know to write:

```
string s("Some text");
```

Most classes go further and have a *default constructor* that initialises the object in a sensible default state. For example, to create an empty vector one simply writes:

```
vector<double> v;
```

The default constructor is automatically called. Similarly, to create an empty string one writes:

```
string s;
```

When you write a class you should almost always give it a default constructor to make it easy for people to use correctly. You may also want to provide more specialised constructors to initialise objects in particular states.

8.7.1   Writing a default constructor

Let us write a class called simply `Point`, which is intended to behave exactly like our `CartesianPoint` example, but which now has a constructor.

```
1  class Point {
2  public:
3      Point(); // declare default constructor
4      double x;
5      double y;
6  };
7
```

```
 8  Point::Point() {
 9      x=0.0;
10      y=0.0;
11  }
12
13  int main() {
14      Point p;
15      cout << "What is the value of x?\n";
16      cout << p.x;
17      return 0;
18  }
```

In the class declaration we now have the declaration `Point();` on line 3. This declares the constructor.

A constructor declaration looks like a function declaration, except:

(i) there is no return type;

(ii) instead of the function name, you have the name of the class.

Similarly the constructor definition looks like a function definition with the same differences. In the example above, the constructor definition begins at line 8.

In our example, the constructor looks like a function that takes no parameters. As we will see, constructors can take parameters. A default constructor is simply a constructor that doesn't take parameters.

You can think of a constructor as a function that is automatically called before anyone is allowed to see the object. Technically speaking it isn't actually a function because it can *only* be called when the object is being initialised and because it doesn't have a return value.

Inside the definition of the constructor you should set all `int`, `double` etc. fields to sensible default values. More generally, you should ensure that the object is in a consistent state before anyone ever sees it and you should perform whatever processing is required to achieve this.

8.7.2 An alternative, and superior syntax

Here is another way that we could have written the constructor for `Point`:

```
class Point {
public:
    Point(); // declare default constructor
    double x;
    double y;
};

Point::Point() :
```

User-Defined Types

```
      x(0.0),
      y(0.0) {
}

int main() {
    Point p;
    cout << "What is the value of x?\n";
    cout << p.x;
    return 0;
}
```

In the definition of the `Point` constructor, you can see a list of statements:

```
      x(0.0),
      y(0.0)
```

This is called an *initialisation list*. This calls constructors for x and y in order to initialise the data. This is slightly different from our earlier code which used an = statement to initialise the variables.

The general syntax for a constructor with an initialisation list is:

```
CLASS_NAME::CLASS_NAME( PARAMETER_LIST) :
    INITIALISATION_LIST {
    ... NORMAL CODE ...
}
```

Notice the colon before the initialisation list.

Experienced C++ programmers prefer an initialisation list because it is marginally faster. This is because calling the default constructor and then performing assignment will be slower than using the right value the first time.

In addition, once you are familiar with the syntax it is more readable because it is immediately obvious that there is nothing clever going on. All that is happening is that data is being initialised. This means that you can guess that if there is some code between the curly brackets it will be relatively interesting.

8.8 Constructors with parameters

Writing constructors that take parameters is straightforward. Here is how one could add a constructor that takes a **strike** and a **maturity** as parameters to the `CallOption` class.

```
class CallOption {
public:
```

```
    double strike;
    double maturity;
    CallOption(); // default constructor
    CallOption(double strike, double maturity);//alternative
};
// default constructor implementation
CallOption::CallOption() :
    strike(0.0),
    maturity(0.0) {
}
// alternative constructor implementation
CallOption::CallOption(
        double s,
        double m ) :
    strike(s),
    maturity(m) {
}
```

To create an option with strike 100 and maturity 2.0 one would then write

```
CallOption myOption( 100, 2.0 );
```

This may seem like a very convenient system, but it is probably not a very good design. The reason is that by introducing classes we got rid of the problem of having to remember in what order to put parameters when calling functions. By having a constructor with multiple parameters, we've reintroduced this problem!

This doesn't mean that constructors that take parameters are always bad, just that you shouldn't over-use them. Certainly the ability to construct a **vector** of given size and default value *is* a good use of constructors.

There is one strange trick with parameterised constructors that you should try to remember. Consider the class **string**. It has a constructor that takes raw text data:

```
string s("Some raw text");
```

C++ uses this constructor to automatically convert raw text data to strings without your having to think about it. For example, the code:

```
plot( "myPlot.txt", xVec, yVec );
```

will work despite the fact that the first parameter of a plot is declared to be a const **string&** and not a **char***.

In general, if your class has a constructor that takes a single parameter construction, C++ will perform similar automatic conversions. While this is great for the **string** data type, usually it results in code that is very unnatural.

For example, suppose you had added a constructor to **BlackScholesModel** where you provide just the stock price:

User-Defined Types

```
class BlackScholesModel {
public:
    double stockPrice;
    double data;
    double volatility;
    double riskFreeRate;
    BlackScholesModel();
    BlackScholesModel( double stockPrice ); // key line
};
```

Doing this means that C++ will now automatically convert doubles into instances of `BlackScholesModel`. This is a very odd thing to do, and is undesirable. To prevent this unwanted behaviour, you should mark constructors that take one parameter as `explicit`.

```
class BlackScholesModel {
public:
    double stockPrice;
    double data;
    double volatility;
    double riskFreeRate;
    BlackScholesModel();
    explicit BlackScholesModel( double stockPrice ); // key line
};
```

The code above prevents automatic conversion. In the very rare event that automatic conversion is useful (as for strings) you can drop the `explicit`.

If you forget the `explicit` it isn't the end of the world. The confusions that may arise from forgetting to use it are quite rare in practice.

Exercises

8.8.1. Write a default constructor for every class that you have created so far. Use both forms of the constructor syntax to familiarise yourself with the options.

8.8.2. Write a class `Polynomial` that represents the polynomial

$$a_0 + a_1 x + a_2 x^2 + \ldots + x^n.$$

Here the coefficients a_i are doubles which associated with the polynomial itself, but x is an unknown.

The class `Polynomial` should have the following features.

144   *C++ for Financial Mathematics*

- It should store the coefficients a_i in a vector.
- It should have a function `evaluate` which takes as a parameter x and evaluates the polynomial at x.
- It should have a constructor which takes a vector of coefficients.
- It should have a function `add` which can be used to add two polynomials.
- It should have a default constructor which generates the constant zero polynomial.
- It should have a constructor which takes a single double c as a parameter and generates the constant polynomial with $a_0 = c$ and all other coefficients 0.

8.9   Summary

- We've learned how to write classes that group data together.
- We've learned how to add member functions to our classes.
- We've seen that using classes makes life easier for users of our library.
- We've learned how to use the keywords `public` and `private` to hide data from users of our library.
- We've learned the buzzword "encapsulation".
- We've learned how to use the `const` keyword to indicate whether or not an object is changed by calling one of its member functions.
- We have learned that classes should normally have a constructor to initialise the data.

Chapter 9

Monte Carlo Pricing in C++

In this chapter we take a break from extending our C++ knowledge and instead work out the details of an in-depth financial example that illustrates everything we have learned so far. With this in mind, let us review some highlights of what we have learned. In the exercises, we have written a large amount of basic mathematical functionality in the file `matlib.cpp`. In particular we can:

- i) Generate random numbers with `randUniform` and `randn` (Exercises 7.7.5 and 7.7.6)

- ii) Generate plots with `plot` and `hist` (Exercises 7.7.8 and 7.7.10).

- iii) Compute statistics of vectors with `min`, `mean`, `prctile`, etc. (Exercises 7.7.2, 7.7.4, and 7.7.9)

We have also learned how to write simple classes.

If you have not yet completed these exercises, you should try to do so. The file `matlib.cpp` which you can find in FMLib9 contains working versions of all these functions.

In this chapter we will do the following:

- i) Add a function to `BlackScholesModel` to simulate stock prices.

- ii) Test our simulations using `mean`, etc., and plot them using `plot`.

- iii) Write a class `MonteCarloPricer` that uses a `BlackScholesModel` to simulate stock prices and then uses risk-neutral pricing to price a `CallOption` by Monte Carlo.

In doing this we will see how all the C++ we have learned can be applied to solve real financial mathematics problems.

Of course, pricing a European call option by Monte Carlo is unnecessary since one already knows the Black–Scholes formula. However, it is a valuable exercise as it illustrates the general ideas of Monte Carlo pricing. In the exercises you will extend the ideas slightly to price a path-dependent option for which we have given no analytic formula.

Before we begin writing code, let us recall the Monte Carlo pricing algorithm which is justified in Appendix A.

145

146   *C++ for Financial Mathematics*

Algorithm 1. *To compute the risk-neutral price of an option whose payoff is given in terms of the prices at times t_1, t_2, ..., t_n one should proceed as follows:*

1. *Simulate stock price paths in the $\mathbb{Q}$-measure. By a stock price path we just mean one possible realisation for the stock prices at the times t_i. A typical stock price path is shown in Figure 9.1.*

2. *Compute the payoff for each price path.*

3. *Compute the discounted mean value. This gives an unbiased estimate of the true risk-neutral price.*

We will proceed in this chapter by first writing a function to simulate stock prices following geometric Brownian motion. This process is described by Equation (A.8). When the drift, μ, of this geometric Brownian motion is set to equal the risk-free rate r, we will be generating stock prices in the risk-neutral measure as described by Equation (A.5). Having written this class, we will write some tests to confirm that our simulations are correct.

Next we will use these stock price paths to compute the price of a European call option by the Monte Carlo method. Again we will write tests to confirm that our answer is correct.

9.1   A function to simulate stock prices

We wish to write a function `generatePricePath` which generates a simulated stock price path. It will take as parameters a final date `toDate` and a number of steps `nSteps`.

Because we want to generate a price path according to a discrete-time version of the Black–Scholes model, we choose to make the function `generate-PricePath` a member function of the class called `BlackScholesModel`. This is logical since the key data that we need to generate the price path are the drift, μ, and volatility, σ. These are already member variables of `BlackScholes-Model`. In addition, by adding the function as a member of `BlackScholes-Model` it becomes immediately clear that the `generatePricePath` function will generate prices that follow geometric Brownian motion, as required by the Black–Scholes model.

This specification for the function effectively tells us what we need to write in the header file.

```
class BlackScholesModel {
public:
    ... other members of BlackScholesModel ...
```

Monte Carlo Pricing in C++

```cpp
    std::vector<double> generatePricePath(
                        double toDate,
                        int nSteps) const;
};
```

As this example shows, the class declaration describes the functionality of the class without describing in detail how it is implemented. If you choose good function and variable names, you won't need too many comments. A well designed class will almost document itself.

We will also want a function `generateRiskNeutralPricePath` which behaves the same, except it uses the $\mathbb{Q}$-measure to compute the path. Therefore we add the following declaration to `BlackScholesModel`

```cpp
class BlackScholesModel {
public:
    ... other members of BlackScholesModel ...

    std::vector<double> generateRiskNeutralPricePath(
                        double toDate,
                        int nSteps) const;
};
```

To implement these functions, we introduce a `private` function that allows you to choose the drift in the simulation of the price path.

```cpp
class BlackScholesModel {
    ... other members of BlackScholesModel ...
private:
    std::vector<double> generateRiskNeutralPricePath(
                        double toDate,
                        int nSteps,
                        double drift) const;
};
```

This function is private because we've only created it to make the implementation easier. Users of the class don't need (or even want) to know about it.

To implement the function we need to know what algorithm to use. The relevant mathematics is given below (and is motivated in Appendix A).

Algorithm 2. *To simulate a stock price that follows discrete-time geometric Brownian motion with drift μ and volatility σ at discrete times t_0, t_1, ...t_n one should proceed as follows:*

(i) Define
$$\delta t_i = t_i - t_{i-1}.$$

(ii) Choose independent normally distributed ϵ_i with mean 0 and standard deviation 1.

148   *C++ for Financial Mathematics*

(iii) Define

$$s_{t_i} = s_{t_{i-1}} + \left(\mu - \frac{1}{2}\sigma^2 \right) \delta t_i + \sigma \sqrt{\delta t_i} \epsilon_i.$$

(iv) Define $S_{t_i} = \exp(s_{t_i})$.

The generated stock prices at S_{t_i} will have the same joint probability distribution as those given by the Black–Scholes model.

Given this mathematical formulation of the algorithm, it is straightforward to implement our helper function. You can find the code in `BlackScholes-Model.cpp`.

```
vector<double> BlackScholesModel::generatePricePath(
        double toDate,
        int nSteps,
        double drift ) const {
    vector<double> path(nSteps,0.0);
    vector<double> epsilon = randn( nSteps );
    double dt = (toDate-date)/nSteps;
    double a = (drift-volatility*volatility*0.5)*dt;
    double b = volatility*sqrt(dt);
    double currentLogS = log( stockPrice );
    for (int i=0; i<nSteps; i++) {
        double dLogS = a + b*epsilon[i];
        double logS = currentLogS + dLogS;
        path[i] = exp( logS );
        currentLogS = logS;
    }
    return path;
}
```

We've only implemented the algorithm for evenly spaced time points. It is easy to generalise to arbitrary time points.

Having implemented the `private` function, the `public` functions are very simple. Again these implementations are placed in `BlackScholesModel.cpp`

```
vector<double> BlackScholesModel::generatePricePath(
        double toDate,
        int nSteps ) const {
    return generatePricePath(toDate, nSteps, drift );
}
```

```
vector<double> BlackScholesModel::
    generateRiskNeutralPricePath(
        double toDate,
        int nSteps ) const {
    return generatePricePath(
```

Monte Carlo Pricing in C++   149

```
        toDate , nSteps , riskFreeRate );
}
```

The point to notice is that by using a design with a `private` helper function, we've avoided writing the same complex code twice.

It is important that we test this code. The first test is a visual one, we want to check that this really looks like a price path. We can use the `LineChart` class to do this.

```
void testVisually () {
    BlackScholesModel bsm;
    bsm.riskFreeRate = 0.05;
    bsm.volatility = 0.1;
    bsm.stockPrice = 100.0;
    bsm.date = 2.0;

    int nSteps = 1000;
    double maturity = 4.0;

    vector<double> path =
        bsm.generatePricePath( maturity , nSteps );
    double dt = (maturity-bsm.date)/nSteps;
    vector<double> times =
        linspace(dt,maturity,nSteps);
    LineChart lineChart;
    lineChart.setTitle("Stock price path");
    lineChart.setSeries(times , path);
    lineChart.writeAsHTML("examplePricePath.html");
}
```

This function makes use of a helper function `linspace` that has been added to the `matlib` library. The call `linspace(a,b,n)` generates n evenly spaced numbers between the values a and b. The output graph is shown in Figure 9.1. The visual check is simply to observe that this looks reasonably like a real stock price's history.

You can run this test using FMLib9 to obtain the relevant code. I hope you agree that producing such a sophisticated chart in C++ is strong evidence of how far we have come in our understanding of C++.

Visual tests are useful, but we should always make sure that we have fully automated tests of our code. The reason is that once one has a large amount of code, it will all need regular testing in case we have accidentally written a bug. If our tests are fully automated, we can run them every time we make any change to the code.

One simple test is that, if our risk-neutral price path function is correct, then the discounted mean of the final stock price should equal the initial price. We test that in the code below:

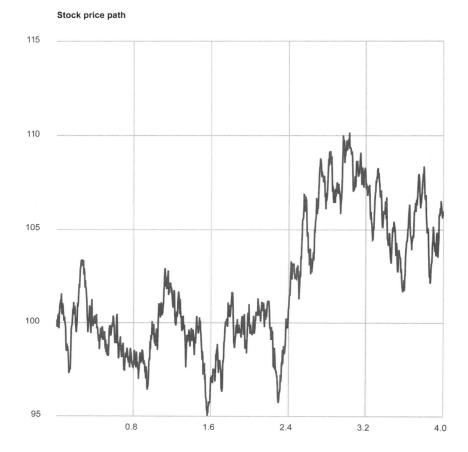

FIGURE 9.1: A simulated stock price path. The bottom axis is the time in years, the vertical axis is the price in dollars.

```
void testRiskNeutralPricePath() {
    rng("default");

    BlackScholesModel bsm;
    bsm.riskFreeRate = 0.05;
    bsm.volatility = 0.1;
    bsm.stockPrice = 100.0;
    bsm.date = 2.0;

    int nPaths = 10000;
    int nsteps = 5;
    double maturity = 4.0;
```

Monte Carlo Pricing in C++

```
    vector<double> finalPrices(nPaths,0.0);
    for (int i=0; i<nPaths; i++) {
        vector<double> path =
            bsm.generateRiskNeutralPricePath(
                maturity, nsteps );
        finalPrices[i] = path.back();
    }
    ASSERT_APPROX_EQUAL( mean( finalPrices ),
        exp( bsm.riskFreeRate*2.0)*bsm.stockPrice,
            0.5);
}
```

We have introduced one new feature here in the line:

```
    rng("default");
```

This is not a call to a standard C++ function, it is a function in the FM-Lib library that resets the state of the random number generator back to its original "default" state. Calls to `randUniform` and `randn` will always generate exactly the same sequence of "random numbers" each time the random number generator is reset to its default state. We have put the phrase "random numbers" in quotes because the random numbers generated are never truly random. Computer-generated random numbers merely *look* random in much the same way as the digits of the number π look random. They are more correctly called *pseudo random numbers*.

Resetting the state of the random number generator is very useful for testing because it guarantees that each time the test is run, we will use the same pseudo random numbers. This means that if the test passes once, it will always pass and similarly if it fails once, it will always fail. It is very frustrating when debugging code to have a test that sometimes passes and sometimes fails—especially if you run all your tests every time you change your code.

Tip: Tests depending on random numbers

Always reset the random number generator to a known state before running tests that involve random numbers.

9.2 Writing a Monte Carlo pricer

We want to write a class called `MonteCarloPricer` that is configured with `nScenarios`, the number of scenarios to generate. This will default to 10000.

152 *C++ for Financial Mathematics*

The `MonteCarloPricer` will also have a function `price` which takes a `CallOption` and a `BlackScholesModel`. It should then compute the price of the `CallOption` by Monte Carlo.

The C++ declaration for `MonteCarloPricer` states pretty much the same thing as the English language specification above.

```cpp
#pragma once

#include "stdafx.h"
#include "CallOption.h"
#include "BlackScholesModel.h"

class MonteCarloPricer {
public:
    /* Constructor */
    MonteCarloPricer();
    /* Number of scenarios */
    int nScenarios;
    /* Price a call option */
    double price( const CallOption& option,
                  const BlackScholesModel& model );
};

void testMonteCarloPricer();
```

Note the now-familiar patterns. The header file is called `MonteCarloPricer.h`. It begins with `#pragma once`. We `#include "stdafx.h"`. The constructor looks like the declaration of a function called `MonteCarloPricer` except there is no return type. We pass the `option` and the `model` by const reference. There is a function to enable testing the code.

The beginning of the C++ file is similarly mundane:

```cpp
#include "MonteCarloPricer.h"

#include "matlib.h"

using namespace std;

MonteCarloPricer::MonteCarloPricer() :
    nScenarios(10000) {
}
```

The C++ file is called `MonteCarloPricer.cpp`. It includes `MonteCarloPricer.h`. The `MonteCarloPricer` has a default constructor which chooses a value for the default number of scenarios to use.

To write the interesting code, we provide an implementation of the `price` function. This follows the Algorithm 1.

```cpp
double MonteCarloPricer::price(
        const CallOption& callOption,
        const BlackScholesModel& model ) {
    double total = 0.0;
    for (int i=0; i<nScenarios; i++) {
        vector<double> path= model.
                    generateRiskNeutralPricePath(
                        callOption.maturity,
                        1 );
        double stockPrice = path.back();
        double payoff=callOption.payoff(stockPrice);
        total+= payoff;
    }
    double mean = total/nScenarios;
    double r = model.riskFreeRate;
    double T = callOption.maturity - model.date;
    return exp(-r*T)*mean;
}
```

Since we're pricing a CallOption whose payoff only depends on the stock price at maturity, we only need one time step for our calculation.

Of course, we also need a test of this functionality. We can compare it with the analytical Black–Scholes formula.

```cpp
static void testPriceCallOption() {
    rng("default");

    CallOption c;
    c.strike = 110;
    c.maturity = 2;

    BlackScholesModel m;
    m.volatility = 0.1;
    m.riskFreeRate = 0.05;
    m.stockPrice = 100.0;
    m.drift = 0.1;
    m.date = 1;

    MonteCarloPricer pricer;
    double price = pricer.price( c, m );
    double expected = c.price( m );
    ASSERT_APPROX_EQUAL( price, expected, 0.1 );
}
```

154   *C++ for Financial Mathematics*

9.3   Generating random numbers for Monte Carlo

As discussed briefly above, conventional computers cannot generate true random numbers, they can only generate *pseudo random numbers*.

Exercise 7.7.5 was to write a function to generate uniformly distributed random numbers using the **rand** function that is built into C++. You should have written some code that looks like this:

```
vector<double> randuniformOld( int n ) {
    vector<double> ret(n, 0.0);
    for (int i=0; i<n; i++) {
        int randInt = rand();
        ret[i] = (randInt + 0.5)/(RAND_MAX+1.0);
    }
    return ret;
}
```

We've called this **randuniformOld** because it uses the old C random number generator **rand**. This isn't a good choice for Monte Carlo algorithms because the sequence of pseudo random numbers it generates start to repeat themselves rather quickly.

Fortunately, experts in generating random numbers on computers have produced much better algorithms for generating pseudo random numbers. The *Mersenne Twister* algorithm is a commonly used random number generator for Monte Carlo simulations which takes advantage of the properties of so-called *Mersenne primes* to generate pseudo random numbers. The details are not important for the purposes of this book. The C++ class **mt19337** allows you to use the Mersenne Twister algorithm without much more difficulty than using the old **rand** function. Admittedly the class name **mt19337** is substantially more intimidating than the old function name **rand** but that is the only big difference. The name comes from the fact that it takes a whopping $2^{19337} - 1$ calls to the random number generator before the sequence of random numbers it generates repeats.

The code to generate random numbers using the **mt19337** class is shown below. It uses several C++ features we have not seen before. First we create a global variable of type **mt19337** called **mersenneTwister**.

```
static mt19937 mersenneTwister;
```

The **mt19937** class is a standard class in modern versions of C++. To use it one has to #include <random>.

The object **mersenneTwister** is our random number generator. We have rewritten the **randuniform** function in **matlib.cpp** so that it uses this random number generator.

Monte Carlo Pricing in C++

```
vector<double> randuniform( int n ) {
    vector<double> ret(n, 0.0);
    for (int i=0; i<n; i++) {
        ret[i] = (mersenneTwister()+0.5)/
                 (mersenneTwister.max()+1.0);
    }
    return ret;
}
```

To generate a random integer we write `mersenneTwister()`. This call returns a random integer in the range `mersenneTwister.min()=0` to `mersenne-Twister.max()`. We've tried to ensure that the values 0 and 1 are never actually achieved, which is why we have added in the `+0.5` and `+1.0`. However, the code is really not very different from the code using the old random number generator.

One new feature is that the class `mt19337` has *overloaded* the meaning of `()`. When we write `mersenneTwister()` it looks as though we are calling a function called `mersenneTwister`. But `mersenneTwister` is an object and not a function, so that can't be right. What is happening is that the `mt19337` has redefined what parentheses mean, so that `mersenneTwister()` means "generate a random number".

You can overload the meaning of most symbols in C++ and this is discussed in detail in Chapter 16. For example, we will show how you can overload `+` and `*` to allow you to perform matrix multiplication.

It is very common in C++ to overload the meaning of `()` so that your objects can be used as though they were functions. This is called writing a *functor* and is discussed in Chapter 19.

We also need a function to reset the state of the random number generator. The code is given below. We simply call seed and pass in the default value for the seed.

```
void rng( const string& description ) {
    ASSERT( description=="default" );
    mersenneTwister.seed(mt19937::default_seed);
}
```

A new feature here is that you can associate data with an entire class. The default value for the seed for the Mersenne Twister algorithm is a constant and so it doesn't make sense to give every instance of the `mt19937` class its own copy of the default value. That would be a waste of computer memory. So instead we associate the default value with the entire class `mt19337`. To access a member variable of a class you use the `::` notation as shown above. We're accessing the variable `default_seed` which is defined on the class `mt19937`. We will discuss this idea further in Section 12.8.

The function `rng` above is designed to work in a similar way as the equivalent function in MATLAB. In general we have tried to make our library `matlib`

156   *C++ for Financial Mathematics*

very similar to the MATLAB library. This is intended to make it easier for users familiar with MATLAB to work with our library.

Note that C++ already contains built-in classes to help with generating random numbers with a given distribution. It obviously makes more sense in practice to use the built-in functions rather than to develop your own. We have written our own `randn` function primarily for educational purposes. To find out about the built-in methods for generating random numbers, see the documentation for the the standard library `<random>`.

Exercises

9.3.1. Add a new function to `MonteCarloPricer` to price a `PutOption`. What aspects of your answer do you find unsatisfactory?

9.3.2. What distribution of prices do you expect at time T according to the Black–Scholes model? Use this to run both visual and automated tests on our generation of stock price paths.

9.3.3. A continuous-time knock-out call option with strike K, barrier B, and maturity T is an option which pays off:

$$\begin{cases} \max\{S_T - K, 0\} & \text{if } S_t < B \text{ for all } t \in [0, T] \\ 0 & \text{otherwise} \end{cases}$$

In other words, it has the same payoff as a call option with strike K unless the stock price hits the barrier level B before maturity, in which case it pays zero (in which one says it has knocked out).

A more practical contract is a discrete-time knock-out call option. This is essentially the same as the continuous-time version, except that you only test whether the stock price is below the barrier at some fixed time points.

Write a class `UpAndOutOption` that represents such an option. Give it a function `computePayoff` that takes a vector of stock prices taken at fixed time points and returns the payoff of the corresponding discrete-time knock-out option.

Add a new function to `MonteCarloPricer` that prices discrete-time knock-out options by simulating price paths of length `nSteps` and computing the payoff for these steps. By taking a large number of steps, the same function can be used to price continuous-time knock-out options.

9.3.4. Write some tests for your answer to the previous question. The challenge is to think of good tests.

Monte Carlo Pricing in C++

9.3.5. To compute the delta, Δ, of an option by Monte Carlo, you can use the following algorithm.

- Choose a small value for h, say $h = S_0 \times 10^{-6}$.
- Generate N stock price paths.
- Use the Monte Carlo method to compute the price of the option when the initial stock price is taken to be $S_0 + h$.
- Use the Monte Carlo method *with the same price paths* to compute the price of the option when the initial stock price is taken to be $S_0 - h$.
- Estimate the delta using the central difference estimate for the derivative
$$f'(x) \approx \frac{f(x+h) - f(x-h)}{2h}.$$

Compute the delta of a call option using this algorithm and write a test for your calculation. You will need to modify the class `BlackScholesModel` a little to implement this.

Explain why it is essential to use the same random numbers in the calculation of the stock prices.

9.3.6. You can use the central limit theorem to estimate the error in our Monte Carlo estimate. In the Monte Carlo method we have generated a sample of N discounted payoffs and taken the mean to compute an estimate for the true expectation. By the central limit theorem, this sample mean is normally distributed about the true expectation with standard deviation $sN^{-1.2}$ where s is the standard deviation of the discounted payoff. We can estimate s by taking the standard deviation of our sample. Use this to compute a 95% confidence interval for the Monte Carlo price.

9.3.7. There are numerous *variance reduction techniques* that can be used to improve Monte Carlo calculations. One approach is called *antithetic sampling*. We first simulate N stock price paths using our $\mathbb{Q}$-measure model Equation (A.4). We recall the formula
$$\log(S_{t+\delta t}) = \log(S_t) + \left(\mu - \frac{1}{2}\sigma^2\right)\delta t + \sigma(\delta t)^{\frac{1}{2}}\epsilon_t.$$

We then simulate N further stock price paths using the same random values ϵ_t but now using the formula
$$\log(S_{t+\delta t}) = \log(S_t) + \left(\mu - \frac{1}{2}\sigma^2\right)\delta t - \sigma(\delta t)^{\frac{1}{2}}\epsilon_t.$$

We can compute an estimate for the risk-neutral price by computing the discounted average payoff for all $2N$ paths. It turns out that for many options this gives a more accurate answer than one obtains by simulating $2N$ independent stock paths. Implement antithetic sampling to price a call option.

You should consult the literature on Monte Carlo methods for more information on this and other variance reduction techniques.

9.4 Summary

Table 9.1 summarises all the files you will find in FMLib9. It states which files you are meant to understand completely and which one you should simply use without understanding.

Financial and mathematical functionality:	
`matlib`	Basic maths functions
`BlackScholesModel`	Represents the Black–Scholes Model
`CallOption`	Represents a call option contract
`PutOption`	Represents a put option contract
`MonteCarloPricer`	Prices options by Monte Carlo
Non-financial functionality:	
`LineChart`	Plots line charts
`Histogram`	Plots histograms
`PieChart`	Plots pie charts
`geometry`	Some elementary mathematical examples
Code you can use, but don't fully understand:	
`testing`	Macros to make testing less boring

TABLE 9.1: Summary of the files found in FMLib9

Consider for a moment what we have achieved. We have demonstrated that by writing code using a number of small classes and functions you can build something very complex out of very little. We have also shown how easy it is to automatically test a large project.

Chapter 10

Interfaces

To complete Exercise 9.3.1 to price a `PutOption` in the last chapter you should have written a function that was simply a copy of the code for a `CallOption` except that the type of the option has changed. This violates the "Once and Only Once" principle (see Section 3.1).

In this chapter we will see how to write a function that can price an infinite number of option contracts by Monte Carlo. It will work equally well with put options, call options, digital put options, digital call options, and even options with payoff functions we've not thought of yet. The possibilities are endless.

10.1 An interface for pricing options

A *path-independent option* is an option whose payoff is completely determined by the stock price at maturity. For example, put options, call options, digital put options, and digital call options are all examples of path-independent options. By contrast, a knock-out option depends not just on the final stock price but upon the entire history of the stock price. This is why a knock-out option is called a *path-dependent option*.

We would like to write a function `price` which takes two parameters:

- a path-independent option,

- a `BlackScholesModel` object,

and which returns the risk-neutral price of the option.

The function will compute the risk-neutral price using the following procedure.

- Compute a large number of possible stock price scenarios at the maturity of the option.

- Compute the option's payoff in each scenario.

- Compute the average payoff and discount by the risk-free rate to get the current price of the option.

159

160   *C++ for Financial Mathematics*

This procedure only works for path-independent options because we only simulate stock prices at maturity and not at intermediate times.

We have just given a recipe for pricing any path-independent option that a human being can easily follow. The question is how to write this recipe in a way that C++ understands. For the time being let us simply guess what the code might look like.

```
double MonteCarloPricer::price(
        const PathIndependentOption& option,
        const BlackScholesModel& model ) {
    double total = 0.0;
    for (int i=0; i<nScenarios; i++) {
        vector<double> path= model.
            generateRiskNeutralPricePath(
                option.getMaturity(),
                1 );
        double stockPrice = path.back();
        double payoff = option.payoff( stockPrice );
        total+= payoff;
    }
    double mean = total/nScenarios;
    double r = model.riskFreeRate;
    double T = option.getMaturity() - model.date;
    return exp(-r*T)*mean;
}
```

The code is essentially identical to the code we wrote to price a `Call-Option`. The only difference is that it takes as a parameter a `Path-IndependentOption`. Of course, the code won't compile just yet. This is because we haven't defined the type `PathIndependentOption`. Just as I had to explain to you what was meant by a path-independent option for you to be able to understand the algorithm, so you have to explain to C++ what is meant by a `PathIndependentOption`.

What the C++ compiler needs to know is that a `PathIndependentOption` is an object which:

- has a fixed maturity;

- has a payoff that depends only upon the stock price at maturity.

We can't just explain this in plain English to the C++ compiler. We have to take a more structured approach. So what we will tell the compiler is that a `PathIndependentOption` is an object which:

- has a function with signature

```
double getMaturity()
```

Interfaces 161

- has a function with signature

```
double getPayoff(double stockPrice)
```

Here we have replaced our plain English description of a path-independent option with a list of functions we expect to be available on any `PathIndependentOption` object. Note that in the implementation of `MonteCarloPricer::price` given above these are the only two functions of the option that we actually use.

This description is called the *interface* of a path-independent option. It describes how we can *use* a path-independent option in our code, but it doesn't describe at all how to *implement* a path-independent option. In fact it doesn't make sense to talk about how to implement any of these functions because there is no one correct implementation. The implementation for a call option and a digital put should be completely different. However, they should both have functions `getMaturity` and `getPayoff`.

You can think of the interface as a *contract* or *specification* that a path-independent option must fulfil. If the class is written in accordance with this specification, we can price it by Monte Carlo.

10.2 Describing an interface in C++

How do we tell C++ that a `PathIndependentOption` is something with a `payoff` and a `getMaturity` function? It's easiest to see with an example. The C++ definition of a `PathIndependentOption` is given below:

```
class PathIndependentOption {
public:
    /*  A virtual destructor */
    virtual ~PathIndependentOption() {}
    /*  Returns the payoff at maturity */
    virtual double payoff(
        double finalStockPrice) const = 0;
    /*  Returns the maturity of the option */
    virtual double getMaturity() const
        = 0;
};
```

The syntax is certainly strange. Your eye is probably drawn to the keyword `virtual` and the use of =0. However, what you should focus on is the similarities with more familiar class definitions.

The class is called `PathIndependentOption`. There is a function called `payoff` and a function called `getMaturity`. They each take and return the correct types. The oddity in their definitions is that they are declared as `virtual` and they are marked with the text `=0`.

The "`=0`" tells C++ that we do not intend to provide a definition for these functions immediately. They have declarations but not definitions. It is impossible to write a `payoff` function that will work for *all* possible types of `PathIndepentOption`, so we tell C++ that we aren't even going to try. We will write other classes with names like `PutOption` and `CallOption` that will each define their own specific payoff function.

In general, when you write an interface, you will not have one particular implementation in mind, so you will write `=0` to mean that you do not intend to provide a definition right now.

The `virtual` keyword is harder to explain. For now, let us just say that it is a magic word you have to utter whenever you write a function without a definition. We'll explain the real meaning of `virtual` in Section 12.3.

You may also be wondering what this strange notation means:

```
/*  A  virtual  destructor  */
virtual  ~PathIndependentOption()  {}
```

For now we'll just pretend that we're writing this code for good luck! It is called a *virtual destructor* and we will explain its meaning once we've got round to saying what a destructor is and what `virtual` actually means. This won't be until Section 16.2. For now, let us simply say that every interface you write should have a virtual destructor and that the syntax for a virtual destructor is:

```
virtual  ~CLASS_NAME()  {}
```

You should carefully notice all the punctuation marks including the ~ character.

Tip: The recipe for writing an interface

- Come up with a suitable name for your class such as `PathIndependent-Option`. This is usually the hard bit. Coming up with a good name for an abstract concept is always tricky.

- Add a *virtual destructor*.

- Add all the methods in your interface. Mark them with `virtual` and `=0`.

At the moment I'm sure you find all this notation perverse and overly complex. You're right. It is. Many computer languages that were invented after C++ have a simpler notation, but in C++ we have to make do with

Interfaces 163

the slightly strange notation. One just has to accept this. We're not learning C++ because it is pretty. We're learning it because many banks program in C++.

Many introductory books on programming in C++ don't mention virtual destructors as early as they should on the grounds that they are difficult to explain. However, they really do need to be there. The philosophy of this book is to show you the right thing to do from the beginning. This is because it's much easier to learn what you should type than why you should type it. Real C++ programmers don't think about every detail of what they type. Instead they have certain habits and make certain default choices. This book aims to teach you good habits sooner rather than later. By way of an analogy, you probably know that it is good practice to say "s'il vous plaît" a lot in French even though you may find it difficult to explain exactly why the French utter this exact combination of syllables or why there is a hat on the letter 'i'.

Now that we have specified our interface, we can tell C++ that our classes `CallOption` and `PutOption` implement the interface. This is done using the following construction:

```
class CallOption : public PathIndependentOption {
```

The line above replaces the usual class declaration. You should read it as saying that a `CallOption` *is a* `PathIndependentOption`. You can think of `: public` as being the C++ expression for is a.

In addition you need to actually implement the functions. Thus we must write declarations and definitions for `payoff` and `getMaturity` in the usual way. In fact we already implemented the `payoff` function. Implementing `get-Maturity` is simple.

With these changes, our `MonteCarloPricer` now works equally well with put options and call options.

```
static void testPutAndCall() {
    rng("default");

    BlackScholesModel m;
    m.volatility = 0.1;
    m.riskFreeRate = 0.05;
    m.stockPrice = 100.0;
    m.drift = 0.1;

    CallOption c;
    c.strike = 110;
    c.maturity = 2;

    PutOption p;
    p.strike = c.strike;
    p.maturity = c.maturity;
```

164   *C++ for Financial Mathematics*

```cpp
    // our pricer can price puts and calls
    MonteCarloPricer pricer;
    double priceC = pricer.price(c, m);
    ASSERT_APPROX_EQUAL(priceC, c.price(m), 0.1);
    double priceP = pricer.price(p, m);
    ASSERT_APPROX_EQUAL(priceP, p.price(m), 0.1);
}
```

You should download FMLib10 now and check that you can compile and run the code. See if you can complete Exercise 10.6.1 at the end of the chapter.

The recipe for implementing an interface

```cpp
class CLASS_NAME : public INTERFACE_NAME {
... declarations for CLASS_NAME ...
};
```

- You will need to provide a new declaration and definition for every function defined in the interface `INTERFACE_NAME`.

- They must have exactly the same types, `const` declarations, and so forth.

10.3   Examples of interfaces

Let us recap:

- We had two classes, `PutOption` and `CallOption`.

- These classes could be used in much the same way, so the code for pricing them was almost identical.

- To avoid duplicating code, we carefully observed which functions `PutOption` and `CallOption` had in common.

- We designed an *interface* which simply listed these common functions, albeit using some unusual syntax.

- We indicated that `PutOption` and `CallOption` implemented the interface.

Interfaces
165

- By telling the C++ compiler about the interface we were able to write one `priceByMonteCarlo` function that works for any path-independent option.

If two different types of objects have similar functions then we can write algorithms that will work with both types of objects by using interfaces. This is a powerful way to avoid duplicating code.

How common is it for two different types of object to have similar functions? The answer is *extremely common*. Here are some examples:

1. Two printers provided by different manufacturers will differ in many ways, but the manufacturers should provide objects that allow us to print files on the different printers without duplicating our code. They can do this by agreeing on a common interface for all printers. This is the same thing as saying they should agree on a common set of print functions all printers are guaranteed to support.

2. Two monsters in a computer game (say an elf and a dragon) will have many functions in common. For example, they will have functions to return their weights, heights, name and to draw them on the screen. The author of the game will introduce an interface `Monster` to avoid duplicating code.

3. All stock price models should have a function `generatePrices`. We can write a standard interface `StockPriceModel` with various implementations such as `BlackScholesModel`, `StudentTModel`, `HestonModel` and `JumpDiffusionModel`. The last three are simply the names of some stock price models that have been proposed in the mathematical finance literature. Our Monte Carlo pricing code can then be modified so it will work with any model. Doing this is left as an exercise, see Exercise 14.4.5. The end result is a very powerful pricing function that can price any path-independent option using any model for the stock price at time T.

4. All trading strategies should have a function `selectPortfolio` which decides how to invest at a particular moment in time. We can have a general interface `TradingStrategy` and concrete subclasses `DeltaHedgingStrategy` and `BuyAndHoldStrategy`. See Exercise 14.4.6.

5. Any encryption algorithm should have methods `encrypt` and `decrypt`. This can be represented as an interface called `Cipher`. It is then possible to provide different implementations of the encryption algorithm that work in different ways. For example, you might write a `CaeserCipher` that replaces `a` with `d`, `b` with `e`, `c` with `f`, etc., or you might write a `ReverseCipher` that replaces `a` with `z` and `b` with `y`, etc. You could then write software that sends encrypted messages and that is capable of working with any `Cipher`.

166 *C++ for Financial Mathematics*

6. Any object of type `GroupElement` should have methods `inverse` and `multiply`. You could then represent any mathematical `Group` in C++ by providing an appropriate implementation of `GroupElement`.

In fact, we actually met the idea of an interface before. An `ostream` is a general data type. We can write data to any stream simply because it has a `<<` function. We don't need to worry about whether it is a file stream or a stream that writes to the screen. What we have been describing in this chapter is how to write your own general types that are as flexible as the `ostream` class.

While the examples above all focus on interfaces in computer software, you can see interfaces in real world design. One example is a car: All cars have steering wheels that behave in the same way. You can drive a car of any make or model because they all make the same basic promise to turn right when you rotate the steering wheel clockwise. Another example is electrical plug sockets. All electrical appliances have the same interface and so they can all be plugged into the same electrical sockets.

Whether we use interfaces for real world design or in software, the payoff is the same. In one example we can learn to drive all cars at once because they have the same interface. In another example, we can price all path-independent options at once because they have the same interface.

We have given a lot of examples because the concept of interfaces is *the* most important idea in object-oriented programming. You might find that at first it is difficult to think of ways in which interfaces could be useful. If so you are not alone. Most programmers find it difficult to take full advantage of interfaces. In practice the only way to learn how to use interfaces is to see lots of examples. In software jargon, examples of how to use interfaces are called "design patterns" after the famous book *Design Patterns: Elements of Reusable Object-Oriented Software* [4]. The aim of this book is to show you the most important design patterns for financial mathematics.

Another piece of software jargon that is useful to know is the word *polymorphism*, which simply means "many forms" in Greek. Polymorphic software is software that can work equally well with a variety of different data types. So a function that uses the `ostream` will be polymorphic because it can be used to write to files, strings, the network, and the screen. Our Monte Carlo pricer is polymorphic because it can be used to price many kinds of options.

10.4 Interfaces in object-oriented programming

In C++ an object consists of two things:

- The data associated with the object,

- All the functions you can use on that object.

A useful mental picture is to think of an object as some data and an instruction manual explaining how to make sense of the data.

When we want to price an option we need to know the strike, the maturity, and also some instructions on how to compute the payoff at maturity. These instructions take the form of a function called `payoff`.

When we call the `price` function on our `MonteCarloPricer` we pass an option object as the first parameter. This object contains all the data we need plus all the instructions we need. When given an option to price, the `price` function looks up the `getMaturity` and the `payoff` functions in the index of the instruction manual, and uses these instructions to perform the computation. Implementing an interface is simply a way of stating clearly that the `getMaturity` and `payoff` functions will be there when needed.

This simple metaphor is the heart of object-oriented programming. When writing software you should pass round not just data, but also the instruction manual for working with that data. An object is nothing but some data and an instruction manual. This is why we pass objects as parameters. This is why we write object-oriented programs.

There are many object-oriented programming languages other than C++. Some other examples that are popular in the financial industry are Java and C#. Object-oriented programming principles are the same in different languages. For this reason a simple system of pictures was invented which allows you to draw the essential features of your software design without worrying about language-specific details. This system of pictures is called UML (Unified Modelling Language). Figure 10.1 shows the UML diagram for our `CallOption`, `PutOption`, and `PathIndependentOption` classes.

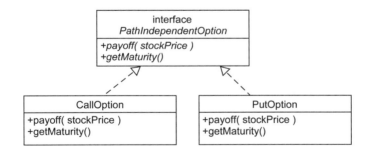

FIGURE 10.1: UML diagram for `PathIndependentOption`

The dotted arrows in Figure 10.1 tell us that the `CallOption` and the `PutOption` classes implement the interface. The italics tell us that `PathIndependentOption` is an interface and that the methods are only declared by `PathIndependentOption` but not defined.

The advantage of the UML diagram is that it strips away some of the more

168   *C++ for Financial Mathematics*

ugly C++ syntax such as the virtual destructor and the `virtual` keyword and lets you see the main points. These main points are:

- All `PathIndependentOption` objects must have a `getMaturity` and a `payoff` function.

- `CallOption` and `PutOption` are two kinds of `PathIndependentOption`.

The UML specification provides an official standard for drawing class diagrams. For example, the + signs are used to indicate that methods are public. Don't be too pedantic about this kind of minor detail when you draw diagrams of your classes. The aim is to produce diagrams that help you summarise what is going on. Just use the parts of UML that you find helpful.

10.5 What's wrong with if statements?

It may have struck you that there is an alternative to using interfaces. Couldn't we write a class such as the class `GeneralOption` given below?

```cpp
class GeneralOption {
public:
    bool isPut;
    bool isDigital;
    double strike;
    double maturity;

    double payoff( double stockPrice );
};
```

You are correct. The payoff function would be a little complex but wouldn't be too difficult to write. This would solve the problem of writing one function that prices puts, calls, digital puts, and digital calls.

However, the real benefit of using interfaces is that our `MonteCarloPricer` can already price *any* path-independent option. It isn't restricted to digital puts and calls. We will be able to use our MonteCarloPricer to price options you've never thought of before: For example, an option whose payoff is given by the tenth root of the stock price.

Even more interesting, observe that the `MonteCarloPricer` code and all the existing option classes will remain completely unchanged even when you think of a new option you want to price. This makes the `MonteCarloPricer` *pluggable*. You can extend its functionality by simply plugging in new options. It is interesting to note once again that electrical plugs are "pluggable" precisely because they have a well-defined interface.

In the context of writing financial software, the use of interfaces means you

Interfaces 169

can extend a trading system with new types of financial contracts, without having to rewrite and retest the whole trading system. Interfaces are the key to solving the complexity issues of writing real financial software.

10.6 An interface for integration

The difficulty with interfaces is not in the syntax or in the concept itself, but in spotting when you can use an interface to solve a problem. For example, we can use interfaces to perform numerical integration, but perhaps the connection between integration and interfaces is not obvious.

If you attempt to write numerical integration code in C++ you will hit a very similar problem to the one we experienced with our Monte Carlo pricer.

For example, you should find it easy to write a function that integrates $\exp(-x^2)$ from a to b using the rectangle rule. But how could you write a function that can integrate *any* real valued function from a to b using the rectangle rule? Interfaces solve this problem.

Let us proceed in the same way as the last example by writing our fantasy code. Here it is:

```
double integral( RealFunction& f,
                 double a,
                 double b,
                 int nPoints ) {
    double h = (b-a)/nPoints;
    double x = a + 0.5*h;
    double total = 0.0;
    for (int i=0; i<nPoints; i++) {
        double y = f.evaluate(x);
        total+=y;
        x+=h;
    }
    return h*total;
}
```

Our fantasy code involves a type called **RealFunction**. This name describes the kinds of object we can integrate. We need to write the formal C++ interface definition.

```
class RealFunction {
public:
    /* A virtual destructor */
    virtual ~RealFunction() {};
    /* This method is abstract, there is
```

170   *C++ for Financial Mathematics*

```
            no definition */
    virtual double evaluate( double x ) = 0;
};
```

This code contains the necessary `virtual` and `=0` magical incantations, but more importantly it states that a `RealFunction` has an `evaluate` method.

Our `integrate` function can integrate any object that provides an `evaluate` method. Here is an example of a specific implementation of our fantasy interface:

```
class SinFunction : public RealFunction {
    double evaluate( double x );
};

double SinFunction::evaluate( double x ) {
    return sin(x);
}
```

The class `SinFunction` represents the mathematical sin function through its `evaluate` method. We can integrate it using our `integrate` function because our `integrate` function uses nothing other than the `evaluate` method.

Combining our `integrate` function, the `RealFunction` interface and its implementation, we can now write the code to integrate `sin`. Here is a test that everything fits together:

```
static void testIntegral() {
    SinFunction integrand;
    double actual = integral(integrand, 1, 3, 1000 );
    double expected = -cos(3.0)+cos(1.0);
    ASSERT_APPROX_EQUAL( actual, expected, 0.000001);
}
```

This code is included in FMLib10. You should check that you can compile and run the test.

Every time we want to integrate a real valued function, we will need to write a new implementation class like `SinFunction`. We don't want to create a new `cpp` file and a new `h` file every time we want to integrate a function, so C++ gives us a shorthand notation that allows us to define a class inside a function. Here is an alternative way of testing the integral of `sin` using a *local class*.

```
static void testIntegralVersion2() {

    class Sin : public RealFunction {
    public:
        double evaluate( double x ) {
            return sin(x);
```

Interfaces 171

```
    }
};

Sin integrand;
double actual = integral(integrand, 1, 3, 1000 );
double expected = -cos(3.0)+cos(1.0);
ASSERT_APPROX_EQUAL(actual, expected, 0.000001);
}
```

Local classes defined in this way are "throwaway". They can only be used inside that one function. The advantage of local classes is that the class is quicker to write. Notice that in the local class declaration we actually provide the definition for all the functions as well as the declaration. Note also that this definition doesn't contain the usual SinFunction:: qualification you need for a standalone class.

Here is a more elaborate example where the local class uses a member variable to access the parameters passed into the containing function.

```
static double integratePayoff(
    double a,
    double b,
    const PathIndependentOption& option) {

    class PayoffFunction : public RealFunction {
    public:
        /* Member variable */
        const PathIndependentOption& option;

        /* Constructor */
        PayoffFunction(
            const PathIndependentOption& option)
            : option( option ) {
        }

        /**
         * Overriding function
         */
        double evaluate(double x) {
            return option.payoff(x);
        }
    };

    PayoffFunction integrand(option);
    return integral(integrand, a, b, 1000);
}
```

172   *C++ for Financial Mathematics*

This function above integrates the payoff function of an option between the range a and b. We have to store the member variable using a reference because you cannot store polymorphic variables by value. Storing data in member variables by reference can be dangerous, because the data referred to may be deleted before the object holding the reference is deleted. This would cause the program to behave badly and perhaps crash. In this case, the code will delete the variable `integrand` from memory the moment `integratePayoff` returns. So there should be no danger of `option` being deleted before `integrand` is deleted.

There is an neater notation called *lambda functions* which can be used to create local classes. We will revisit the example of writing an integrator in Chapter 19 and show how the use of lambda classes simplifies the code.

We should also add that the rectangle rule is not the best integration method to use. You probably already know of Simpson's rule. An even more interesting technique for one-dimensional numerical integration is called Gaussian quadrature. Numerical libraries such as the GNU scientific library contain sophisticated integration routines that you should use in practice.

Exercises

10.6.1. Write classes `DigitalCallOption` and `DigitalPutOption`. Price them using the `MonteCarloPricer`. See Appendix A if you do not recall what a digital option is.

10.6.2. Write a class `NormalPDF` and compute its integral from -1.96 to 1.96.

10.6.3. Write a function `integralToInfinity` that uses integration by substitution to transform an infinite integral to one the function `integral` can compute.

10.6.4. Test the `normcdf` function against numerical integration of the pdf of the normal distribution.

10.6.5. Write a function `differentiateNumerically` that can differentiate any `RealFunction`.

10.6.6. Write an interface `Cipher` and provide at least two implementations of your choice. Write a single function `void testCipher(Cipher& toTest)` that can test any Cipher works correctly.

10.6.7. Draw a UML diagram for your `Cipher` classes.

10.6.8. Write a class `RectangleRulePricer` which can price a `PathDependentOption` using the numerical integration routines written in the earlier exercises. Use Equation (A.3) to compute the probability density function

Interfaces 173

for the log of the stock price. Check your solution works with a `CallOption`. This gives a numerical test of the derivation of Equation (A.6).

10.6.9. Design an interface called `ContinuousTimeOption`. It should have a function `payoff` which takes a vector of stock prices and a vector of times at which the price occurred. This should return the payoff that would have occurred if the stock price had varied linearly between the time points. (Obviously we can't actually supply the stock price at an infinite number of points). The `ContinuousTimeOption` should also have a function called `getMaturity()`.

Write an implementation of `ContinuousTimeOption` called `KnockOutCallOption`. Add a function to `MonteCarloPricer` which can approximate the price of any `ContinuousTimeOption` given a finite number of time steps.

10.6.10. Show that you could use `if` statement to write an `integrate` function which can numerically integrate any of the following specific functions (but no others):

- $\sin(x)$
- $\cos(x)$
- $\exp(x)$

Explain why this is not a good design. Use this to explain why using an `if` statement to decide how to price puts and calls differently is not as powerful as using an interface.

10.7 Summary

- By defining interfaces we can write code that will work with inputs we haven't thought of yet.

- This is essential to writing real code for the financial industry where new products are invented every day.

- Interfaces are common in everyday life. Cars, doors, guitars, and plug sockets are all familiar examples of things that are easy to use because they implement a well-known interface.

- Learning how to use interfaces is the most important skill in object-oriented design.

Chapter 11

Arrays, Strings, and Pointers

One useful way to view C++ is as a collection of languages:

(i) the C language;

(ii) an object-oriented programming language;

(iii) a template programming language.

This chapter is about some of the C language aspects of C++. We have ignored them so far, because you should focus your efforts on the object-oriented features of C++. Similarly we will not discuss the template programming features until Chapter 17.

Pointers are a central feature of the C language. They are a tool for directly manipulating computer memory. You will not find them in a more modern pure object-oriented programming language like Java. The feature is omitted from other languages because pointer-based programs are typically harder to understand than object-oriented programs.

Most of what you learn in this chapter you should never use directly. The only thing you need to use is `shared_ptr`. We will discuss this towards the end of the chapter. Unfortunately, to understand `shared_ptr` you must first understand ordinary pointers.

There are three other motivations for studying the material in this chapter.

(i) The first is that C++ was originally designed for people with C programming experience. So some C++ code is designed to be intuitive to use if you have experience with C pointers. If you don't study pointers, you will find things like the C++ standard library very unintuitive.

(ii) The second is that job interviewers may assume that you know C and so may ask some trick questions about pointers, so it is good to be prepared!

(iii) The third is that to write maximally efficient C++ programs it can be helpful to manage memory consciously. Pointers are a tool for direct memory manipulation.

Because this chapter and the next are more about C programming than object-oriented programming, we've temporarily abandoned developing FM-Lib for this chapter. Instead the project `PointerExamples.zip` contains all code for this chapter.

175

176 *C++ for Financial Mathematics*

As you will see, there is just one large `main.cpp` file which contains all the various examples in sequence. The `main` method just calls various test functions in sequence. Some of the tests have been commented out because they illustrate what you *shouldn't* do. If you include them, the program will probably crash.

11.1 Arrays, the C alternative to `vector`

The code below shows how you create an array consisting of 5 integers in C++. An array is a bit like a vector in that it contains a sequence of values, but as we will see, arrays are not as useful as vectors.

```
// Create an uninitialised of length 5
int myArray[5];
for (int i=0; i<5; i++) {
    cout<<"Entry "<<i<<"=";
    cout<<myArray[i];
    cout<<"\n";
}
```

This code creates an array of 5 integers, without initialising it. It then runs through the entries and prints them out. Since we haven't initialised the values, it will print out some random integers.

Note the similarities with vector. Just as with vector, we use [] to access the entries. Just as with vector, the entries start at 0.

The one difference you can spot is that we are not using the `size()` function. This is because a C array has no size function. Actually, you can use the `sizeof` operator and write `sizeof(myArray)/sizeof(int)` where we have written 5. This would be better programming style, but I've avoided it because it is less readable when you are new to arrays.

To create an array that is initialised one proceeds as follows:

```
// Create an initialised array
int myArray[] = {1, 1, 2, 3, 5};
for (int i=0; i<5; i++) {
    cout<<"Entry "<<i<<"=";
    cout<<myArray[i];
    cout<<"\n";
}
```

Simply specify the values in a comma-separated list between curly brackets. Notice that if we provide a list of values, we no longer have to specify the length of the array when we create it.

Arrays, Strings, and Pointers 177

Its quite natural to want to initialise an array so that all its values are zero. Here is the relevant code.

```cpp
// Create an initialised array
int myArray[5] = {};
for (int i=0; i<5; i++) {
    cout<<"Entry "<<i<<"=";
    cout<<myArray[i];
    cout<<"\n";
}
```

Here we specify the size of the array and write ={} to mean "set everything to zero". Actually, this is just a special case of a more general method of initialising an array.

```cpp
int myArray[5] = {1,2,3};
for (int i=0; i<5; i++) {
    cout<<"Entry "<<i<<"=";
    cout<<myArray[i];
    cout<<"\n";
}
```

Here we specify the length and the first few values. The rest of the values are padded with zeros. So the code above will print out the values 1, 2, 3, 0, 0.

One problem with arrays is that because they have no `size` function we must pass their length as well as their value when we pass them to another function.

```cpp
int sumArray( int toSum[], int length ) {
    int sum = 0;
    for (int i=0; i<length; i++) {
        sum+=toSum[i];
    }
    return sum;
}
```

Unfortunately, the `sizeof` trick I mentioned earlier stops working the moment you pass arrays to other functions. So you really do need to pass the length as well as the array.

Although we can pass arrays to functions, you must *never return them from a function*. When you return other data types such as `double` or a `vector` the value is copied when you return it. But this doesn't happen for arrays. You can think of them as being returned by reference.

This causes problems, because when a function returns, all the variables it has created are removed from memory. If you attempt to return an array, the caller of your function will just be given a pointer to the memory address where the array *used* to be. The computer may have reused that memory for almost anything. As a result, the behaviour of the program becomes completely unpredictable. Typically it will crash, sometimes horribly. The same happens if you try to return a local variable by reference, but you're less likely to try doing that.

If you like you can create multi-dimensional arrays. The syntax should be self-explanatory. The only strange thing is that you have to specify all the dimensions but the first by hand.

```
// Create an initialised 3x5 array
int myArray[][5] = {{1, 2, 3, 4, 5},
                    {2, 0, 0, 0, 0},
                    {3, 0, 0, 0, 0}};
for (int i=0; i<3; i++) {
    for (int j=0; j<5; j++) {
        cout<<"Entry ("<<i<<","<<j<<")=";
        cout<<myArray[i][j];
        cout<<"\n";
    }
}
```

Let us not waste any more time understanding arrays in more detail, because we will never use them!

Arrays are not very useful data structures for us because *you cannot change the length of an array.* You can't insert a new item or add some items at the end. Even worse, the size of an array is fixed at compile time. This means when you write your program, you must choose the size of the array and it can't grow in response to user input. Also, passing the length around to each function call is annoying.

Arrays are only of interest if you know the size of your arrays when your code is compiled. For example, when working with 3D computer graphics, arrays of length 3 can be useful.

Tip: Don't use arrays

Vectors do everything you want to do with arrays and much more.

Arrays, Strings, and Pointers 179

11.2 Pointers

11.2.1 new and delete

Because arrays are not flexible enough, the C language also contains the concept of a pointer. This allows a C programmer to work with sequences of data of varying lengths.

```
int n = 5;
int* myArray = new int[n];
for (int i=0; i<n; i++) {
    cout<<"Entry "<<i<<"=";
    cout << myArray[i];
    cout << "\n";
}
delete[] myArray;
```

The code above uses the **new** ... [] operator to allocate a chunk of memory. In this case we are creating a sequence of `int` data types in memory, but you could use other data types instead.

The good news is that you can choose the size of the chunk of memory at run time. Unlike with an array, you don't have to choose the size in advance when you write your program.

The downside of using **new** [] is that the memory will *not* be automatically deleted when the variable goes out of scope (see Section 4.7 for the meaning of the word scope). You must use `delete` [] to free the memory created with the **new**[] operator. This is good and bad. With arrays we couldn't return arrays because the memory was deleted automatically, but on the other hand we didn't have to remember to call `delete[]`. With memory created using **new**[] we have to remember to delete the memory by hand, but you can safely return the data.

The value returned by **new** `int[n]` is called a *pointer*. It has type `int*` which means "a pointer to an `int`". In actual fact the array discussed in the previous section was of type `int*` too. It's just that the notation for arrays in C hides the fact that an array-valued variable is just a pointer to a memory location. All that the `int* myArray` contains is the memory address where the array starts. As a result, we have to remember ourselves that the block of memory is of length **n**.

You can use [] with a pointer to find the integer at a given offset in just the same way as we use [] with arrays and vectors. We have used this in the example above to print out the values in the array.

If you run the code example above, you will find that the entries in the array take arbitrary values. Just as local variables of type `int` have unpredictable initial values, so does `int` data created with **new**.

180   *C++ for Financial Mathematics*

You can use **new[]** with other data types too. To illustrate this let's introduce a simple class, **Pair**.

```
class Pair {
public:
    double x;
    double y;
    Pair();
    Pair( double x, double y );
};
```

We assume that the default constructor of **Pair** initialises x and y to zero. Given this class, the code below creates a chunk of memory containing pairs. This time each pair is initialised to $(0,0)$ since **Pair** has a default constructor.

```
    int n = 5;
    Pair* myPairs = new Pair[n];
    for (int i=0; i<n; i++) {
        double xValue = myPairs[i].x;
        double yValue = myPairs[i].y;

        cout<<"Pair (";
        cout<< xValue;
        cout<<",";
        cout<< yValue;
        cout<<")\n";
    }
    delete[] myPairs;
```

The type is a **Pair***. This should be read "as a pointer to a **Pair**".

11.2.2   Pointer operators

A pointer is just the address in memory of some data. On a 32-bit computer a pointer will be 4 bytes long. On a 64-bit computer it will be 8 bytes long. This means a 32-bit computer can have up to 2^{32} bytes $\approx$ 2Gb of memory. 64-bit computers could in principle have vastly more memory. It is normal to write memory addresses as a hexadecimal number; e.g., **9ABF0132** is a typical memory address.

Consider the following code:

```
    int* fivePrimes = new int[5];
    fivePrimes[0] = 2;
    fivePrimes[1] = 3;
    fivePrimes[2] = 5;
    fivePrimes[3] = 7;
    fivePrimes[4] = 11;
```

After running the code, there will be an area of memory that looks as shown in in Figure 11.1. In this figure each box denotes one `int`-sized unit of memory. The x's mark memory location whose value we know nothing about. What we do know is that there are five consecutive entries containing the numbers 2, 3, 5, 7, and 11. The pointer `fivePrimes` contains the memory address of the first of these numbers.

FIGURE 11.1: A pointer to a memory location containing an array of the first five primes.

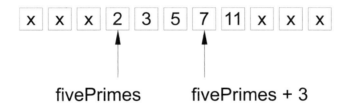

FIGURE 11.2: Pointer arithmetic.

You can extract the memory address of any variable even if it was not created with **new**. You use the operator **&** to do this. You can then store this memory address in another variable so long as that variable is declared to have a pointer type. Given a pointer, you can access what it points to using `*`. These ideas are illustrated in the following code:

```
int myVariable = 10;
int* pointerToMyVariable = &myVariable;

cout << "Memory location of myVariable ";
cout << pointerToMyVariable;
cout << "\n";

cout << "Value of myVariable ";
cout << (*pointerToMyVariable);
cout << "\n";
```

182   *C++ for Financial Mathematics*

I ran the code above on a 32-bit computer and got the following output:

```
Memory location of myVariable 0013FE78
Value of myVariable 10
```

Notice that `pointerToMyVariable` is a pointer to an `int`, so it is of type `int*`. Since `myVariable` is an `int`, `&myVariable` is a pointer to an `int`.

Since `pointerToMyVariable` is a pointer to an `int`, `*pointerToMyVariable` is an `int`.

It is worth emphasising that we use `&` in two completely different ways. We've used this symbol previously for pass by reference. We're using it now to convert a variable into a pointer. These uses are completely unconnected. It just happens to be the same character. When used as part of a type definition, `&` means "reference". When used as an operator, `&` means "memory location of".

We also use `*` in three different ways. We use it in type definitions to mean "this is a pointer". We use it to convert pointers back to references. And (in case you'd forgotten) we also use it for multiplication!

Another very useful operator for working with pointers is `->`. You can use `->` to access the fields of an object via a pointer.

```cpp
    Pair p;
    Pair* pointerToP = &p;

    // Use -> to access fields via a pointer
    pointerToP->x = 123.0;
    pointerToP->y = 456.0;

    // We check that c has changed
    ASSERT( p.x==123.0 );
    ASSERT( p.y==456.0 );

    // You could use * and .
    ASSERT( (*pointerToP).x==123.0 );
    ASSERT( (*pointerToP).y==456.0 );
```

Using `->` is equivalent to using `*` and `.` in combination but is easier to read.

You can also add and subtract integers to a pointer. For example, if `p` is a pointer to a memory location, then `p+1` is a pointer to the next memory location, and `p+3` is a pointer to three memory locations along. This is illustrated in Figure 11.2. To add one to a pointer, you can write `p++`. Working with pointers in this way is called *pointer arithmetic*.

11.2.3   Looping with pointers

We can pass a pointer to a function much as we pass an array to a function. Indeed, there isn't really any difference between a function that takes an array

Arrays, Strings, and Pointers 183

as a parameter and a function that takes a pointer as a parameter. Here is a function that adds up a sequence of `length` numbers which start at the memory location `toSum`.

```
int sumUsingPointer( int* toSum, int length ) {
    int sum = 0;
    for (int i=0; i<length; i++) {
        sum+=toSum[i];
    }
    return sum;
}
```

The code here is identical to that with arrays except that we declare the type using * rather than []. In particular, notice that you have to pass the length as well as the pointer.

There isn't a lot wrong with this code. Nevertheless, an experienced C++ programmer might choose to write this as follows:

```
int sumUsingForAndPlusPlus(int* begin, int n) {
    int sum = 0;
    int* end = begin + n;
    for (int* ptr = begin; ptr != end; ptr++) {
        sum += *ptr;
    }
    return sum;
}
```

This code is hard to follow if you are new to C++. But because it follows certain programming conventions, it does become easy to understand with a little practice.

We call the first memory location `begin`. We call the memory location immediately after the last element `end`. In the `for` loop we move the variable `ptr` through the values from `begin` up to, but not including, `end`. At each stage of the loop we add to `sum` the value pointed to by `ptr`.

There are two reasons why an experienced programmer might prefer this version of the code:

(i) The code is completely standardised and therefore easy to read once you are used to it. `begin` is a standard name for a pointer as is `end`. This means that the reader can understand the `for` loop at a glance. This is the purpose of for loops: they shout out "boring looping code" so you can focus on the interesting bit in the middle of the loop and ignore the rest. Later in the book we will discuss a C++ concept called iterators that allows this pointer code to be generalised to many different kinds of data structures. See Chapter 18.

(ii) This code may be slightly faster. We only perform one addition per loop, namely `ptr++`. The original code contains the addition `i++` and implicit addition when we write `toSum[i]`. To compute this value we need to add `i` to the pointer `toSum` and then read the memory address. You can see this more clearly if we rewrite the first version of the code using pointer arithmetic instead of `[]`.

```
int sumUsingPointerArithmetic( int* toSum,
                               int length ) {
    int sum = 0;
    for (int i=0; i<length; i++) {
        int* ithElement = toSum + i;
        int valueOfIthElement = *ithElement;
        sum+= valueOfIthElement;
    }
    return sum;
}
```

Tip: Avoid writing brainteasers

Used in moderation, pointer arithmetic can be very helpful. For example, we will take advantage of pointer arithmetic when we write a class to represent a matrix in Chapter 16. However, you should try to avoid writing code that leaves people scratching their heads. Excessive use of pointers and pointer arithmetic is an excellent way to write incomprehensible code.

As an example, notice that a memory location may itself contain a pointer. A pointer to a pointer to an `int` is of type `int**`. There is no reason to stop there if you like confusing yourself and your colleagues. There is nothing *technically* wrong with an `int****`. Here is a genuine job interview question that exploits this.

Question: What would be the equivalent pointer expression for referring to the array element `a[i][j][k][l]`?

A. `((((a+i)+j)+k)+l)`

B. `*(*(*(*(a+i)+j)+k)+l)`

C. `(((a+i)+j)+k+l)`

D. `*((a+i)+j+k+l)`

Answer: The correct answer, if you still want the job, is B. The correct answer in practice is to avoid this kind of silly code.

Arrays, Strings, and Pointers 185

11.2.4 Using pointers in practice

Since we always need the number of elements as well as the pointer, it seems wise to introduce a class such as:

```
class IntArray {
public:
    int* firstElement;
    int  length;
};
```

It also seems a good idea to give this class a helpful function `size`.

Before we go any further, we should stop and realise that this class already exists. It is a `vector<int>`.

Thus although we've now painfully introduced pointers, we could do everything we wanted with `vector` already!

Tip: Avoid new []

You should avoid using `new` `[]` and simply work with `vectors` instead. The only possible exception might be if you believe that you are such a good programmer you will be able to get a bit more performance out of accessing raw memory. This is unlikely to be true in practice.

11.3 Pointers to text

In C you don't have classes, and this includes the class `string`. As a result, in C the standard convention is to represent text using a block of memory containing characters terminated by the character code 0.

Recall that a `char` is just an 8-bit number. The standard encoding is ASCII where "A" is encoded as 65, the symbol "1" is encoded as 49, and the symbol "0" is 48. The code number 0 on the other hand isn't used for any character symbols, so it can be safely used to mark the end of a block of memory. You use a backslash to create the special character with code number 0 as demonstrated in the code below.

```
    char charArray1[] =
        {'H', 'e', 'l', 'l', 'o', '\0' };
    for (int i=0; i<6; i++) {
        cout << "ASCII VALUE ";
        char c = charArray1[i];
        cout << ((int)c);
```

FIGURE 11.3: A pointer to a memory location containing the null-terminated string Hello.

```
        cout << "\n";
}
```

C provides a shortcut for creating a sequence of characters ending with the zero code, just place the desired characters in double quotes. You don't need to include the zero at the end, this is included automatically.

```
const char* charArray2 = "Hello";
for (int i=0; i<6; i++) {
    cout << "ASCII VALUE ";
    char c = charArray2[i];
    cout << ((int)c);
    cout << "\n";
}
```

This format of data is called a "C-style string" or a "null-terminated string". It is illustrated graphically in Figure 11.3. Note that we have said that the type of charArray2 is const char*. This means that content of the string is fixed.

Earlier in the book we've created strings by writing the text in quotes, but we've tried to avoid the char* data type as far as possible. This is because you should always try to convert a char* into a string as soon as possible.

Since this wasn't possible in C, C contains various functions to help work with null-terminated strings, such as strlen, which computes the length (excluding the terminating zero character) and strcpy, which copies one string into another.

The code below shows how to use pointers directly to replicate the behaviour of the strlen function.

```
int computeLengthOfString( const char* s ) {
    int length = 0;
    while ((*s)!=0) {
        s++;
        length++;
    }
    return length;
```

Arrays, Strings, and Pointers
187

```
}

void testComputeLengthOfString() {
    const char* quotation="To be or not to be";
    int l1 = computeLengthOfString( quotation );
    int l2 = strlen( quotation ); // built in
    ASSERT( l1==l2 );
}
```

When working with arrays and segments of memory, you must be very careful not to attempt to access data outside the bounds of your array. For example, the code below will behave unpredictably and will probably crash horribly.

```
char* shortText = new char[20];
for (int i=0; i<1000; i++) {
    shortText[i] = 'x';
}
delete[] shortText;
```

If you have downloaded the project `PointerExamples`, you can run this test by uncommenting the `testOverrunning` test in the main method. I expect that you'll get a pretty nasty error message.

When you use classes such as `string` and `vector` in debug mode, various additional checks are included so that you get more helpful error messages. You could try rewriting this code using a `vector<char>` and compare the error messages obtained. This is another very compelling reason for using classes rather than pointers directly.

Tip: Avoid char*

`string` is designed to get rid of the many problems programmers have historically had with `char*`. If you are performing serious string manipulations, for example if you're writing a web server, even `string` is not up to the job as it can't cope with international characters.

11.4 Pass by pointer

We have already met the ideas of *pass by value* and *pass by reference* (see Section 7.2.1). In principle you can also pass data to a function using *pass by pointer*.

```
void polarToCartesian(double r, double theta,
```

188 *C++ for Financial Mathematics*

```
    double* x, double* y) {
    *x = r*cos(theta);
    *y = r*sin(theta);
}

void testPolarToCartesian() {
    double r = 2;
    double theta = atan(1);
    double x;
    double y;
    polarToCartesian(r, theta, &x, &y);
    ASSERT_APPROX_EQUAL(x, sqrt(2), 0.0001);
    ASSERT_APPROX_EQUAL(y, sqrt(2), 0.0001);
}
```

This code passes the parameters r and theta by value in the usual way, but passes pointers to the parameters x and y. These are local variables of the test function. When polarToCartesian is called it directly changes these local variables. It is able to do this because it has been passed the memory addresses where the variables x and y are stored.

There is very little practical difference between passing data using a pointer and passing data using a reference. In the C language, you have to use pass by pointer because the concept of reference does not exist.

In the C++ language, using pass by reference is the preferred approach.

The most practical difference is that the user of the function polarToCartesian is forced to use the & symbol to obtain pointers to the variables x and y. This is, in some regards, a good thing because it makes it clear that x and y might be modified by the function call. On the other hand, using pointers does add a layer of conceptual difficulty.

The only real difference in what you can achieve using pointers and references is that if you pass a pointer it is not entirely clear whether or not it is acceptable to use pointer operators such as incrementing the pointer with ++ or calling delete. On the other hand, if you use pass by reference it is clear that the object won't be deleted and the ++ operator won't be called because you simply cannot apply these operators to a reference. This is why pass by reference is the preferred approach in C++.

Just as with references, you can use the keyword const with a pointer. There are two possible ways to combine the keyword const with a pointer. Writing const on the right of the * means that the pointer itself cannot be modified. Writing const on the left of the * means it is *the data that is pointed to* that cannot be modified rather than the pointer itself. This distinction is illustrated in the code below. If you remove the comment characters, the code will not compile.

```
void constPointerExamples() {
    const char* ptr = "A string";
```

Arrays, Strings, and Pointers

189

```cpp
    ptr++;
    std::cout << (*ptr);
    // (*ptr)='a';  // You can't change the data
    // using a const char*

    char* fiveChars = new char[5];
    char *const constPtr = fiveChars;
    // constPtr++;  // You can't change a char *const
    (*constPtr) = 'h';  // but you can change
                        // what it points to
}
```

If you find the difference between a `const char*` and a `char *const` confusing, then you are not alone. One might argue that the conceptual difficulties introduced by using `const` outweigh the benefits of compile time checking. As we have said before, the lead developer on a project should have a view on this as one should use `const` consistently or avoid using it altogether.

If you are planning on using `const`, then this discussion gives one other reason for preferring to pass by reference to pass by pointer: There is only one way to use `const` with references, so that's one less thing to think about. Note that this might help you remember where to put the `const` when working with pointers. `const` on the left of the `*` means the same as `const` on the left of `&`. You can't even write `const` on the right of `&` because you can't use operations like `++` with references.

11.5   Don't return pointers to local variables

Just as with arrays, you have to make sure that when you return a pointer, you aren't returning a pointer to a local variable that is about to be deleted. For example, this code is incorrect:

```cpp
char* thisFunctionReturnsAnArray() {
    /* This produces a compiler warning */
    char text[] = "Don't do this";
    return text;
}

void someOtherFunction() {
    char text[] = "Alternative text\n";
    cout << text;
    cout << "\n";
}
```

```
void testDontReturnArrays () {
    char* text = thisFunctionReturnsAnArray ();
    someOtherFunction ();
    cout << text;
    cout << "\n";
}
```

The text here is a char array so it will be deleted the moment the function exits. This means that the code will behave unpredictably. It will probably print some junk if you run it.

You can return a string without a problem because it is returned by value: A new copy of the string is given to the caller. The same applies to returning a vector.

You are allowed to return a pointer to data created with new []. However, you will then have to make sure the caller knows whether or not they will be expected to call delete[] at some point.

By convention in C and C++, if a function returns a pointer the caller is *not* expected to call delete[]. For example, if you call c_str() on a string you shouldn't delete[] what it returns. When you call c_str it creates a temporary C-style string in memory that may be deleted when the string is modified or when the string is deleted. This is the standard recommended pattern in C++. If you create something it is your job to delete it.

As an example, the code below and the test is technically valid but violates the convention that you shouldn't delete a pointer returned by another function. As such most C++ programmers would agree that this is poor code.

```
char* thisFunctionReturnsAPointer () {
    char text[] = "This works";
    int n = strlen(text);
    char* ret = new char[n+1];
    /* We now get a compiler warning here */
    strcpy( ret, text );
    return ret;
}

void testReturnPointerJustAboutOK () {
    char* text = thisFunctionReturnsAPointer ();
    someOtherFunction ();
    cout << text;
    cout << "\n";
    // don't forget to free the memory
    delete[] text;
}
```

> **Tip: Returning pointers**
>
> Never return a pointer to a local variable. If you wish, you may return pointers to member variables of an object. In this case it is the object's job to make sure the data is deleted rather than the job of the caller. (See Section 16.2 for information on how to achieve this using *destructors*)

11.6 Using pointers to share data

Here is an example of the type of problem we now wish to solve:

Example 1: We have a class `Instrument`. It contains lots of data about a traded instrument. For example, the type of the instrument, the Bloomberg code, the Reuter's code, etc. You should imagine that each `Instrument` takes up a considerable amount of memory.

We have another class, `Position`, which consists simply of an instrument, the quantity held in that instrument, and the name of the trader who has taken the position.

We wish to save memory by reusing the same `Instrument` instances. This is illustrated in Figure 11.4.

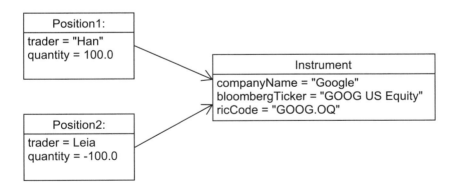

FIGURE 11.4: Two `Position` objects sharing an `Instrument`.

The data we have used so far has all been stored in local variables or member variables.

Local variables are no good for long-term storage because their data is deleted the moment the variable goes out of scope.

Member variables only last as long as the object containing the variable. So data stored in member variables will be lost as soon as the object is deleted.

Using `new`, we can create long-term storage for data. This data will not be automatically deleted, so we have to manually delete it with a call to `delete`. `new` and `delete` are just like `new[]` and `delete[]` except that they create and delete single objects rather than arrays. You have to be very careful to use `delete[]` when you have used `new []` and `delete` when you have used `new`.

Here is an example of how to create individual pair instances using `new`. Note the call to `delete` and the use of `->`.

```
Pair* myPair = new Pair;
myPair->x = 1.3;
myPair->y = 2.5;

cout << "Pair (";
cout << (myPair->x);
cout << ", ";
cout << (myPair->y);
cout << ")\n";

delete myPair;
```

Let us see how to use data created in this way to share `Instrument` data between two positions.

First we define classes for `Position` and `Instrument`:

```
class Instrument {
public:
    string bloombergTicker;
    string ricCode;
    string companyName;
    Instrument() {}
};

class Position {
public:
    string trader;
    double quantity;
    Instrument* instrument;
    explicit Position( Instrument * instrument );
};
```

Arrays, Strings, and Pointers 193

```
Position::Position( Instrument* instrument ) :
    instrument( instrument ) {
}
```

You should notice that the `Position` contains a pointer to an Instrument and not an actual Instrument. You can see this because the type of the field Instrument is `Instrument*`.

The code below initialises a vector of `Position` objects that both point to the same `Instrument`. This achieves our goal of sharing data.

```
vector<Position> constructPositions() {
    // the caller of this function
    // should call delete on the instrument
    // when they are done with all the positions
    vector<Position> positions;

    Instrument* instrument = new Instrument;
    instrument->companyName = "Google";
    instrument->bloombergTicker = "GOOG US Equity";
    instrument->ricCode = "GOOG.OQ";

    Position p1(instrument);
    p1.trader = "Han";
    p1.quantity = 100.00;
    positions.push_back( p1 );

    Position p2(instrument);
    p2.trader = "Leia";
    p2.quantity = -100.00;
    p2.instrument = instrument;
    positions.push_back( p2 );

    return positions;
}
```

However, there is a catch, the caller of this function has been given the task of deleting the instruments.

```
void testConstructPositions() {
    vector<Position> r = constructPositions();
    int n = r.size();
    for (int i=0; i<n; i++) {
        cout << "Position "<<i<<"\n";
        Position& p=r[i];
        cout << "Trader "<<p.trader<<"\n";
        cout << "Quantity "<<p.quantity<<"\n";
        cout << "Instrument ";
```

C++ for Financial Mathematics

```
        cout << p.instrument->companyName <<"\n";
        cout << "\n";
    }
    delete r[0].instrument;
}
```

The caller of `constructPositions` has to know precisely how to call `delete`. This violates information hiding—you need to know how `construct-Positions` actually works to be able to use it safely.

Danger!

Initialise your pointers. We have initialised the field `instrument` in the constructor. If you don't initialise a pointer, it will fail horribly when you try to use it.

```
// Don't do this
Instrument* instrument;
cout << instrument->companyName << "\n";
```

Sometimes you wish to specify that a pointer doesn't yet point to anything, in which case you initialise it to `nullptr`. Here is code that checks to see if a pointer is `nullptr` before using it.

```
string getCompanyName( Position& position ) {
    if (position.instrument==nullptr) {
        return "Name not set";
    } else {
        return position.instrument->companyName;
    }
}
```

The danger in using `nullptr` is that if you forget the check you will get a nasty error.

11.6.1 Sharing with shared_ptr

The class `shared_ptr` solves the `delete` problem. To use `shared_ptr`, you need to `#include <memory>` to use `shared_ptr`.

`shared_ptr` is an example of a so-called "smart pointer". This is a C++ class which behaves like a pointer but which handles working out when to call `delete` on your behalf. In fact, `shared_ptr` is the most versatile smart pointer class and the only one we will use in this book.

The way a `shared_ptr` works is that each shared pointer keeps track of how often it has been copied. Once the number of copies of the smart pointer in

Arrays, Strings, and Pointers

existence drops to zero, the data pointed to will be deleted for you. How this is actually achieved will have to wait until we discuss destructors in Section 16.2.

The changes one makes to use a `shared_ptr` are very simple. Here is a new version of the `Position` class which uses a `shared_ptr`.

```
class PositionV2 {
public:
    string trader;
    double quantity;
    shared_ptr<Instrument> instrument;
    explicit PositionV2(shared_ptr<Instrument> ins );
};

PositionV2::PositionV2(shared_ptr<Instrument> ins ) :
    instrument( ins ) {
}
```

The difference is that it now contains a `shared_ptr<Instrument>` rather than an `Instrument*`. The former means "shared pointer to an instrument", the latter means "pointer to an instrument". The former is actually easier to read, although it is undeniably harder to type.

When we use a `shared_ptr` we can use the operators `->` and `*` just as if we were working with a pointer.

```
vector<PositionV2> constructPositionsV2() {
    vector<PositionV2> positions;

    shared_ptr<Instrument> ins
        = make_shared<Instrument>();
    ins->companyName = "Google";
    ins->bloombergTicker = "GOOG US Equity";
    ins->ricCode = "GOOG.OQ";

    PositionV2 p1(ins);
    p1.trader = "Han";
    p1.quantity = 100.00;
    positions.push_back( p1 );

    PositionV2 p2(ins);
    p2.trader = "Leia";
    p2.quantity = -100.00;
    p2.instrument = ins;
    positions.push_back( p2 );

    return positions;
}
```

196 *C++ for Financial Mathematics*

The key difference is that we use the function `make_shared` instead of `new` to create a `shared_ptr`.

```
shared_ptr<Instrument> ins
    = make_shared<Instrument>();
```

`make_shared` is defined in `<memory>` and is part of the namespace `std`.

The payoff of introducing `shared_ptr` can be seen in the calling function.

```cpp
void testConstructPositionsV2() {
    vector<PositionV2> r = constructPositionsV2();
    int n = r.size();
    for (int i=0; i<n; i++) {
        cout << "Position "<<i<<"\n";
        PositionV2& p=r[i];
        cout << "Trader "<<p.trader<<"\n";
        cout << "Quantity "<<p.quantity<<"\n";
        cout << "Instrument ";
        cout << p.instrument->companyName<<"\n";
        cout << "\n";
    }
}
```

The essential point is that we no longer need to call `delete`, so information hiding has been restored.

Danger!

Programming with raw pointers is hard. You may feel that you will never forget to call `delete`, and so you can use pointers without using `shared_ptr`, thereby writing faster code. If so, here is an example of a problem that might catch you out.

```cpp
int* a = new int[10];
int* b = new int[20000000];
cout << "Arrays created";
delete[] a;
delete[] b;
```

What's wrong with this code? Well, what happens if the second call to `new` fails because we are running short of memory and so throws an error? Then the first object will never be deleted. So your program has just leaked memory when it was already running low on memory. It will probably crash soon.

`shared_ptr` works very well for most practical purposes but it does have the problem that if you have circular references between your data objects,

Arrays, Strings, and Pointers 197

then they will never get deleted. The cure for this is either to avoid circular references or to use an even smarter pointer library. In this book, we choose to simply avoid circular references.

11.7 Sharing data with references

A lot of what you can do with a pointer, you can do with a reference. For example, you can have a member variable which is a reference. If you store data by reference you save memory just as you do if you store data by pointer.

One advantage of using references rather than pointers is that you *must* initialise member variables which are references in the constructor. This means that the errors you can get from null pointers and uninitialised pointers just can't happen with a reference.

However, using reference member variables is not normally as useful as choosing `shared_ptr` member variables. The reason is that owning a `shared_ptr` to an object means that you are guaranteed the object won't be deleted until you no longer need it. On the other hand, if you use a reference, there's a danger that someone else might delete your object.

Let's give an example of why you shouldn't normally use references as member variables. First we define a new `Position` class that contains a reference to an `Instrument`.

```
class PositionV3 {
public:
    string trader;
    double quantity;
    Instrument& instrument;
    explicit PositionV3( Instrument& instrument );
};

PositionV3::PositionV3( Instrument& instrument ) :
    instrument(instrument) {
}
```

This code is technically correct. Notice in particular that we initialise a member variable reference in the constructor as is required.

The problem with this design is that the `Position` class has no control over when the `instrument` is deleted. The following code seems to work at first, it certainly compiles:

```
PositionV3 constructPositionV3() {
    // This function doesn't work, the instrument
    // is deleted, so all the returned positions
```

198   *C++ for Financial Mathematics*

```
    // contain broken references
    vector<PositionV3> positions;

    Instrument instrument;
    instrument.companyName = "Google";
    instrument.bloombergTicker = "GOOG US Equity";
    instrument.ricCode = "GOOG.OQ";

    PositionV3 position(instrument);
    position.trader = "Han";
    position.quantity = 100.00;
    return position;
}
```

The problem with this code is that we're creating a `Position` that appears to be initialised but then immediately deleting the `Instrument` it refers to. We're not consciously deleting the instrument, it just happens as part of automatic cleanup when the method exits. This means that the following code which uses the returned position will fail.

```
void testConstructPositionV3() {
    // This will fail horribly
    PositionV3 p = constructPositionV3();
    cout << "Trader "<<p.trader<<"\n";
    cout << "Quantity "<<p.quantity<<"\n";
    cout << "Instrument ";
    cout << p.instrument.companyName<<"\n";
    cout << "\n";
}
```

You can run this code by uncommenting the appropriate line in the `main` method. It will crash.

This doesn't mean that using reference member variables is completely forbidden. Indeed, you should already have done so for one of the exercises in the previous chapter. Here is a C++ function which transforms an infinite integral into a bounded integral and then integrates it.

```
double integralToInfinity(RealFunction& f,
    double lowerLimit, int nPoints) {

    class DefiniteIntegrand : public RealFunction {
    public:
        RealFunction& g;
        double lowerLimit;

        DefiniteIntegrand(RealFunction& g,
                          double lowerLimit) :
```

Arrays, Strings, and Pointers 199

```cpp
        g(g), lowerLimit(lowerLimit) {
    }

    double evaluate(double x) {
        return (1/(x*x))
            * g.evaluate(lowerLimit - 1 + (1/x));
    }
};

DefiniteIntegrand integrand(f, lowerLimit);
return integral(integrand, 0, 1, nPoints);
}
```

This code is perfectly good C++ because we know that our `Definite-Integrand` instance will only be kept in memory until the `integralToInfinity` method returns. We know that the function passed in to `integralToInfinity` will not be deleted until `integralToInfinity` returns. Therefore we can safely store a reference to it as a member variable.

Tip: Think carefully before using reference member variables or returning references

You can return references and you can store data by reference but you will need to be certain when the reference will be deleted. You should never return a reference to a local variable as that will always be deleted. You will need to think carefully when returning references to member variables.

Thinking carefully takes energy, so if you don't have any to spare use a `shared_ptr`.

11.8 The C++ memory model

One of the main features of C and C++ that make it appealing to computer programmers is that you can manipulate computer memory directly. For financial mathematicians this is often more of an irritation than a blessing, since computer memory isn't their particular area of expertise. However, if you wish to use pointers effectively and to debug code that uses pointers, it helps to understand computer memory.

You can think of computer memory as divided into four sections:

(i) The memory used by other programs. You are not allowed to access this memory. If you try to then the operating system will make your program

200   *C++ for Financial Mathematics*

crash. This kind of bug is called a *segmentation fault* or a *general protection fault*. You can easily produce such an error in C++ by initialising a pointer to a random hex value and then trying to access memory using that pointer.

(ii) The memory containing the computer code for your program. You can read from this memory but can't write to it. If you try to write to this memory, your program will crash.

(iii) A region of memory called the *stack*. This is used for short-term storage for calculations. You can think of the stack as rather like a stack of paper on which rough notes are made and then quickly discarded. The local variables you use in a function are stored on the stack. We will describe the stack in Section 11.8.1.

(iv) A region of memory called the *heap*. This is used for longer-term storage for larger data structures. You can think of it as being a bit like a filing cabinet containing records you expect you will need to refer to repeatedly. Data created using **new** is stored on the heap. We will describe the stack in Section 11.8.2.

11.8.1   The stack

Consider the following program which tests a `factorial` function.

```cpp
int factorial(int n) {
    int ret = 1;
    for (int i = 0; i<n; i++) {
        ret *= (i + 1);
    }
    return ret;
}

void testFactorial() {
    int n = 3;
    int nFactorial = factorial(n);
    assert(nFactorial == 6);
}

int main() {
    testFactorial();
    return 0;
}
```

The function **main** calls a function **testFactorial** which in turn calls the function **factorial**. While this function executes, all data is stored on the *stack*. You can see this because there are no calls to **new** and we don't call

Arrays, Strings, and Pointers   201

any functions that might be calling `new` without us knowing about it (except possibly `assert`).

When line 6 is executed the stack will look as shown in Figure 11.5:

factorial ⎰	int n	
	int ret	
	Called by `testFactorial` at line 11.	
testFactorial ⎰	int n	
	int nFactorial	
	Called by `main` at line 16	
main {	Called by the operating system.	

FIGURE 11.5: A snapshot of the stack.

For each function being executed, the stack stores all the parameters and local variables. It also contains a reference to where that function was called. The computer needs a record of where the code was called in order to be able to move the processing back to the correct location on return.

The moment `testFactorial` returns, the memory used by that function can be deleted. The local variables `n` and `ret` of `testFactorial` are simply discarded from the stack. The stack will now look as shown in Figure 11.6.

testFactorial ⎰	int n
	int nFactorial
	Called by `main` at line 16
main {	Called by the operating system.

FIGURE 11.6: A snapshot of the stack once the factorial function has returned.

This simple system of using a single block of memory for all function calls and simply adding and removing data at the top as new functions are called and then return is extremely efficient. It is one of the reasons why C++ code can be particularly fast.

However, the stack is not sufficient to solve all problems. For example, the data allocated on the stack is always of a fixed size that is known when the program is compiled. This is why C arrays are of a fixed size that is determined before the program is compiled. As we have discussed, this makes C arrays too inflexible for most purposes and so it is usually better to use the `vector` class instead. Although it isn't immediately apparent to the end user, a `vector` must store its data on the heap.

11.8.2 The heap

The heap is an area of memory that is less organised than the stack. As shown in Figure 11.7, the stack is a neatly arranged block of memory, whereas the heap contains objects arranged in a haphazard manner.

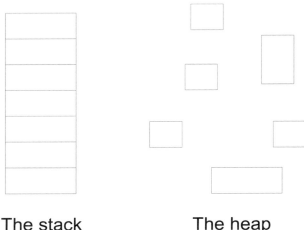

FIGURE 11.7: The stack and the heap

When you call `new`, a clever memory management algorithm hunts for an unused area of the heap that is large enough for your requirements. When you call `delete`, that same memory management algorithm is informed that it can now reuse that area of computer memory.

The big advantage of the heap is that you can call `new` and `delete` whenever you like and you can choose the size of the memory allocated at run time rather than at compile time.

Many mathematical problems can be solved without using the heap consciously at all. When you create a data object such as a `string` or a `vector` it will create data on the heap on your behalf, but you usually don't need to think about this in detail. However, there are times when you will need to use flexibility of the heap yourself.

The key example for us will be when pricing a portfolio of options. We don't expect to know when the program is compiled exactly what our portfolio will look like. We don't know how many options we will hold or even what types of option we will hold. Since the memory used by the stack is fixed at compile time, it cannot be used to store an unknown portfolio of options. In Chapter 13 we will show how to write a Portfolio class which uses `shared_ptr` to store the data of a Portfolio on the heap.

Arrays, Strings, and Pointers 203

Bugs with pointers
Using pointers introduces a number of new kinds of potential bug into our code.

- *A null pointer error.* This is where you try to access data at the memory address `nullptr` or `0`.

- *A general protection fault (GP), also known as a segmentation violation (SEGV).* This is where you try and access memory you aren't allowed to access because you are using an invalid pointer.

- *A memory leak.* This is where you call `new` but never call `delete`. Memory leaks tend not to be obvious immediately. They usually show up as a program that gradually starts running more and more slowly to the point where you have to close it and reopen it. Using `shared_ptr` is the best way to avoid memory leaks. There are also tools such as `INSURE++` that will help you detect memory leaks.

These are extremely common problems in C++ code and you are sure to experience them at some time.

Exercises

11.8.1. Write a function `sumDoubles` which takes a pointer to a list of doubles and the length of the list and returns the total. Write a test for this function.

11.8.2. Write a function `reverseDoubles` which takes a pointer to a list of doubles and the length of the list and reverses it. So $(1, 2, 3, 4)$ should be changed to $(4, 3, 2, 1)$, for example. Write a test for this function.

11.8.3. Write a function `meanDistance` which takes a pointer to a list of `Pair` objects and computes the mean distance of (x, y) to the origin.

11.8.4. Write a function `polarToCartesian` which takes four parameters: `r`, `theta`, `x`, and `y`. `x` and `y` should be passed by pointers. The function should populate the values pointed to by `x` and `y` with the $r\cos(\theta)$ and $r\sin(\theta)$. Write a test for this function.

In the C language, one does not have references, so passing by pointer is the C equivalent of passing by reference.

11.8.5. Write a function `reverseString` which takes a `char*` string and reverses it. This should be similar to your `reverseDoubles` function except you need to consider the terminating zero. Write a test for your function.

204   *C++ for Financial Mathematics*

11.8.6. Write a function `concatenate` to append one null-terminated string to another. I've deliberately not told you the parameters or return type, this is a design question for you to solve. When you have come up with your own solution, take a look at the standard library functions `strcat` and `strcat_s` to see how others have solved this problem. Of course, you should use the library functions rather than write your own.

11.8.7. Write a function `search` which takes as input two `char*` strings. The first should be `phrase`, a phrase to search for, the second should be `text`, some text to scan through. It should return the number of times `phrase` appears in `text`. So `search("be","to be or not to be")` should return `"2"`.

11.9   Summary

Pointers are an essential tool for a C programmer, but aren't so important to a C++ programmer. Nevertheless you must use pointers to some extent if you want to work with data of variable sizes or with long-lived data. By far the easiest way to use pointers is to use the `shared_ptr` class.

- Pointers, references, and `shared_ptr` can be used to achieve similar things.

- Use `vector` and `string` rather than arrays or `char *`.

- References are a bit safer than pointers, for example you can't call `delete` on a reference and you *must* initialise reference variables. So pass by reference is preferred to pass by pointer.

- A `shared_ptr` is a bit slower than a reference or a pointer, but is really the only viable option for long-lived data if you want to avoid memory leaks.

Chapter 12

More Sophisticated Classes

In this chapter we discuss a variety of additional features that C++ provides to allow you to write more complex and interesting classes.

12.1 Inlining member functions

We discussed in Section 5.4.4 that you can use the keyword `inline` with a function, as a hint to the compiler that you want the machine code for the function to be copied every time it is used. This has the advantage of being slightly faster because calling a method uses a little computer power. For example, we chose to inline the functions `hornerFunction` used in the definition of `norminv`.

You can also inline member functions of a class. The procedure is to simply place the definition inside the class in place of the declaration.

```
class Point {
public:
    double getX() const {
        return x;
    }
    // other members of Point
private:
    double x;
    double y;
};
```

In this example, `getX` is inlined. This means that using this function will result in code that is no slower than code that accesses `x` directly. Using a method has the advantage that if we change our mind about how a Point stores its data we can do so without needing to rewrite the code that uses `Point`.

You will probably have noticed that writing an inline function is much less tedious than writing a separate declaration and definition. Unfortunately it isn't advisable to use inlining functions just to avoid typing. You should write separate definitions only if you actually want the function to be inlined.

205

12.2 The this keyword

In order to hide information, it is a good idea to have no `public` variables. We have just seen how to write a `get` method to provide read access to a variable. In the example below we show how to write a `set` method to provide write access.

```cpp
class Point {
public:
    double getX() const {
        return x;
    }
    void setX(double x) {
        this->x = x;
    }
    // other members of Point
private:
    double x;
    double y;
};
```

We have used the new keyword `this`. When implementing a member function the variable `this` always contains a pointer to the current object. So you can use `this->` to access the fields of the current object. We've just demonstrated this with the set method.

You can also use `*this` to obtain a reference to the current object. To see why you might want to do this, suppose that we want to add a `price` method to a class `UpAndOutOption`. We decide that to implement this, we'll simply pass the necessary data to our existing `MonteCarloPricer` class and ask it to perform the calculation. However, the class `MonteCarloPricer` requires you to pass it an option object. Where do we find a reference to the appropriate option object? `*this` is just what we need to pass to the pricer.

```cpp
double UpAndOutOption::price(
        const BlackScholesModel& model ) const {
    MonteCarloPricer pricer;
    return pricer.price( *this, model );
}
```

12.3 Inheritance

Inheritance is a powerful technique for reusing code in C++. It is a technique specifically designed to make it easier to implement interfaces. To give an example of how inheritance can be useful, let us consider an example of an interface which at first sight seems quite time consuming to implement.

The following interface defines a `ContinuousTimeOption`.

```cpp
class ContinuousTimeOption {
public:
    /*  Virtual destructor */
    virtual ~ContinuousTimeOption () {};
    /*  The maturity of the option */
    virtual double getMaturity () const = 0;
    /*  Calculate the payoff of the option given
        a history of prices */
    virtual double payoff (
        const std::vector<double>& stockPrices
        ) const = 0;
    /*  Is the option path-dependent?*/
    virtual bool isPathDependent () const = 0;
};
```

A continuous time option has a maturity and we can compute the payoff come maturity by looking at the price path up to maturity. This is the idea captured by the `ContinuousTimeOption` interface.

There are three functions to implement in `ContinuousTimeOption`. Most of the implementation will be the same for a `PutOption`, a `CallOption`, a `DigitalPutOption`, etc. *Inheritance* gives a way of implementing methods for a number of classes at once.

To use inheritance, one begins by defining a so-called "base class" that defines functions we wish to reuse. We will call our base class `Continuous-TimeOptionBase`. Here is a declaration of an appropriate base class. We're using the inlining technique here to implement appropriate get and set methods.

```cpp
class ContinuousTimeOptionBase :
    public ContinuousTimeOption {
public:
    virtual ~ContinuousTimeOptionBase () {}
    double getMaturity () const {
        return maturity;
    }
    void setMaturity (double maturity) {
        this->maturity = maturity;
```

208   *C++ for Financial Mathematics*

```
    }
    double getStrike() const {
        return strike;
    }
    void setStrike(double strike) {
        this->strike = strike;
    }
    // ... more methods ...
private:
    double maturity;
    double strike;
};
```

Notice that our base class has a virtual destructor.

```
    virtual ~ContinuousTimeOptionBase() {}
```

We've seen these before with interface classes. Just like interface classes, any class you use as a base class must have a virtual destructor.

We can now write a class `PutOption` which "extends" the base class. It will automatically "inherit" all the functions.

```
class PutOption : public ContinuousTimeOptionBase {
public:

    /* Calculate the payoff of the option given
       a history of prices */
    double payoff(
        const std::vector<double>& stockPrices
        ) const;

    double price( const BlackScholesModel& bsm )
        const;

    bool isPathDependent() const {
        return false;
    };
};
```

The first line of the code says that the class `PutOption` extends the class ContinuousTimeOptionBase.

```
class PutOption : public ContinuousTimeOptionBase {
```

This means that all the methods that are defined in `ContinuousTimeOption-Base` are automatically provided for `PutOption` so we don't need to write them again.

More Sophisticated Classes 209

In particular, `PutOption` inherits the functions `getMaturity`, `getStrike`, etc., and the data `maturity` and `strike`. There is no need to redeclare or redefine any of these.

In addition, `PutOption` inherits the interface `ContinuousTimeOption` from `ContinuousTimeOptionBase`. So a `PutOption` will automatically be an instance of `ContinuousTimeOption`.

```
class ContinuousTimeOptionBase :
    public ContinuousTimeOption {
```

Notice that we use the same notation in C++ to say that we extend a class or that we implement an interface.

12.3.1 What have we gained?

So far we seem to have only made things more complicated. But we can now easily write the following classes:

- `CallOption`,

- `DigitalCallOption`,

- `DigitalPutOption`,

- `UpAndOutOption`.

If all of them extend `ContinuousTimeOptionBase`, we won't have to write `getMaturity`, `getStrike`, etc., methods for any of them.

This gets rid of a tremendous amount of repetitive code. Notice that the very idea of interfaces, with a fixed set of methods that must be implemented by all classes, naturally leads to somewhat repetitive code. You should use inheritance to avoid this kind of code repetition.

12.3.2 Terminology

Various terms are used to describe the relationship between `PutOption` and `ContinuousTimeOptionBase`.

- `ContinuousTimeOptionBase` is termed a *superclass* or a *parent class* of `PutOption`.

- `PutOption` is termed a *subclass* or a *child class* of `ContinuousTime-OptionBase`.

- `PutOption` *extends* from `ContinuousTimeOptionBase`.

- `PutOption` *inherits* from `ContinuousTimeOptionBase`.

12.4 Overriding methods — the `virtual` keyword

Let us add a new method to `ContinuousTimeOptionBase` to price an option given the pricing model.

```
class ContinuousTimeOptionBase
    : public ContinuousTimeOption {
public:
  /* Price the option, by Monte Carlo or otherwise*/
  double price(
      const BlackScholesModel& model) const;
  // ... other members ...
};
```

To implement the method using Monte Carlo we would simply use the following code:

```
double ContinuousTimeOptionBase::price(
      const BlackScholesModel& model ) const {
    MonteCarloPricer pricer;
    return pricer.price( *this, model );
}
```

This implementation works reasonably for all option types, but it isn't necessarily optimal. For example, a `PutOption` is best priced analytically rather than by Monte Carlo.

Therefore we would like to be able to *override* the default implementation. The keyword `virtual` means that a function may be overridden in a subclass. So we must add it to the declaration of `price`

```
class ContinuousTimeOptionBase
    : public ContinuousTimeOption {
public:
  /* Price the option, by Monte Carlo or otherwise*/
  virtual double price(
      const BlackScholesModel& model) const;
  // ... other members ...
};
```

If you wish to override a method in a subclass you should do the following:

(i) Check it is declared as `virtual` in the superclass.

(ii) Declare it with precisely the same parameter and return types (and uses of `const`) in your subclass. Optionally, you can use the keyword `override` so the compiler checks that you have done this correctly.

More Sophisticated Classes 211

(iii) You must provide the new implementation that overrides the default behaviour.

Here is an example for the PutOption class.

```
class PutOption : public ContinuousTimeOptionBase {
public:
    double price(const BlackScholesModel& bsm)
        const override;
    // ... other members ...
};
```

This means that the PutOption has a different price function than the default one. The implementation of this method is just the usual Black–Scholes pricing formula that we've seen before.

The override keyword does not affect the behaviour of the program when it is running. It is simply a way of saying that your intention is to override a function. If you do not use this keyword and you make a small error in the list of parameter types of your function then the compiler may not notice your error. This can lead to bugs that are hard to spot. Therefore it is advisable to use override whenever you wish to override a function.

When you call price on an option, the appropriate price method will be automatically selected based on the type of the option. This will allow us to write a class that prices an entire Portfolio of options in the next chapter.

12.4.1   A note on the keyword virtual

The keyword virtual means "may be overridden".

When you write an interface class, the methods have no definitions, so we must always declare them as virtual because they *must* be overridden.

All classes have a destructor. This is a special function called when the object is destroyed. This must be declared as virtual in any class that may be subclassed so that the overridden destructor will be called.

This explains why we had to use the keyword virtual several times in the definition of an interface class.

Danger!

Virtual destructors: any class that is designed to be subclassed should have a virtual destructor. Unfortunately most compilers don't pick up on it if you forget to write one.

212 *C++ for Financial Mathematics*

12.5   Abstract functions =0

You can say that a function has no implementation by writing =0. This simply means that there is no default implementation, the subclass must provide its own implementation. Such functions are called abstract functions. Clearly, functions without definitions must be declared as `virtual` so that they can be overridden.

An interface class is a class which has *only* abstract functions. As we have shown you can write classes which have a mixture of implemented and non-implemented functions. An abstract class is a class that contains at least one abstract function. For example, `ContinuousTimeOptionBase` is an abstract class because it inherits the abstract `payoff` function but doesn't implement it.

12.6   Multiple layers

We aren't restricted to just parent classes and child classes. We can build complex hierarchies of classes.

For example, a class can have a grandparent. Indeed, we've already seen this, `PutOption` has the parent `ContinuousTimeOptionBase` and grandparent `ContinuousTimeOption`.

You can continue ad infinitum. For example we could (and should) introduce a new class, `PathIndependentOption`, which contains the code common to puts and calls but that is not used by knock-out options and Asian options.

Here is an appropriate declaration for a `PathIndependentOption`:

```cpp
class PathIndependentOption :
        public ContinuousTimeOptionBase {
public:
  /* A virtual destructor */
  virtual ~PathIndependentOption() {}
  /* Returns the payoff at maturity */
  virtual double payoff(double endStockPrice) const
      = 0;
  /* Compute the payoff from a price path */
  double payoff(
    const std::vector<double>& stockPrices ) const {
    return payoff(stockPrices.back());
  }
  /* Is the option path dependent? */
```

```
    bool isPathDependent() const {
        return false;
    };
};
```

The implementation of `PathIndependentOption` does the following.

- It provides an implementation of `isPathDependent`.

- It has an abstract function to compute the payoff given only the final stock price. By definition of "path independent" a path-independent option should be able to provide an implementation of this function.

- This means we can implement the `payoff` function that takes an entire path of stock prices. We only need to look at the last price.

Once we've written a `PathIndependentOption` class, we can use this to write a `CallOption` class that extends it.

```
class CallOption : public PathIndependentOption {
public:

    double payoff( double stockAtMaturity ) const;

    double price( const BlackScholesModel& bsm )
        const;
};
```

To implement a `CallOption` we will have to write a function to compute the payoff at maturity. So the `payoff` function above must be declared.

We have also chosen to provide an override for `price`. This is because we know that it would be better to use the Black–Scholes formula to price a Call Option than to use the default Monte Carlo implementation.

Notice that all the functions `getMaturity`, `getStrike`, etc., are implemented by the grandparent `ContinuousTimeOptionBase`.

Similarly, the `ContinuousTimeOption` method `payoff` is implemented by the parent `PathIndependentOption`, so we only have to implement a simpler method that computes the payoff using the price at maturity.

We could write `PutOption`, `DigitalCallOption`, and `DigitalPutOption` in a similar way with very little additional code. This is left as an exercise.

12.6.1 UML

Figure 12.1 shows a UML diagram for the class hierarchy of options in FMLib12. Actually, the `DigitalPutOption` and `DigitalCallOption` don't exist. You should write them yourself as an exercise.

The arrows in Figure 12.1 (with the specific shape of arrow head shown)

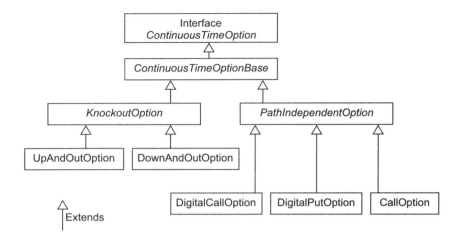

FIGURE 12.1: The hierarchy of option classes.

mean "extends". The boxes denote classes. UML uses other types of arrows to indicate other relationships between classes. We will see examples later. We have followed the standard practice in UML of using italics to indicate abstract classes and roman type to indicate concrete classes.

This UML diagram is unusually complex. Class hierarchies are usually much simpler. We have tried to simplify the diagram a little by omitting all details about the member functions and variables of our classes.

Our complex hierarchy is justified because it represents the "true" relationship between options. `CallOptions` really are `PathIndependentOptions` which really are `ContinuousTimeOptions`. This is a pattern you should follow in general. You should only use subclassing to mean "is a". All the arrows in our UML diagram can be read as "is a".

> **What good software looks like**
>
> Which artist do you prefer, Mark Rothko or Hieronymus Bosch? If you are not familiar with their paintings, they are both great painters and are polar opposites in their style. Rothko is known for abstract paintings which have a stunning simplicity. Bosch is known for his detailed landscapes of fantastic imagery.
>
> When it comes to art, you don't actually have to pick a favourite. But in software design you should always prefer the simplest solution to a problem.

More Sophisticated Classes 215

> This means that when you design your classes you should aim for a UML diagram that looks more like a Rothko than a Bosch.

12.6.2 Another object hierarchy

To give another example hierarchy, the `geometry` library has been updated to contain an interesting object hierarchy. It contains the following.

- A class `Shape` that represents any finite shape in the plane. It has the following methods.

    - A method `area` to compute the area.
    - A method `contains` to test if a point is in the shape.
    - A method `boundingRectangle` that returns a `Rectangle` containing the entire shape.

- The class `Circle` is one implementation of `Shape`.

- The class `Rectangle` is another implementation of `Shape`.

- The class `HyperCircle` (the shape $x^4 + y^4 < 1$) is another implementation of `Shape`.

- The class `Shape` has a default implementation for `area` that uses Monte Carlo.

- `Circle` and `Rectangle` override area.

These classes would be useful in a graphics library.

Notice the importance of abstract functions and overriding. No one would want a graphics library whose `Shape` class couldn't be extended. Similarly no one would want a pricing library where you couldn't define new types of options.

These examples strike at the heart of object-oriented programming. Abstract methods and overriding methods allow you to write "pluggable" systems.

12.6.3 Multiple inheritance

In `C++` a class can extend more than one base class. This is called *multiple inheritance*.

Multiple inheritance is useful to indicate that a class implements more than one interface.

As an example, it might be useful to have an interface `DerivativeWithStrike` for derivatives that have some associated strike price such as call options and put options (but not futures). The interface would simply return the strike price as shown below:

216   *C++ for Financial Mathematics*

```
class DerivativeWithStrike {
public:
    ~DerivativeWithStrike();
    virtual double getStrike() const = 0;
};
```

To declare that a `ContinuousTimeOptionBase` implements both `Continuous-TimeOption` and `DerivativeWithStrike`, one would write:

```
class ContinuousTimeOptionBase :
    public ContinuousTimeOption,
    public DerivativeWithStrike {
```

When using multiple inheritance, I recommend that you ensure that at most one of the parent classes is a non-interface class. This is not a rule that is enforced by `C++`, but if you break it you will have to understand the detailed rules of multiple inheritance in `C++`. This makes your code harder to understand and therefore harder to maintain.

Tip: Multiple inheritance

Only use multiple inheritance to show that a class implements more than one interface.

We have only shown how to publicly extend from another class. This is why we use the keyword `public` whenever we extend a class. It is also possible to extend a class privately, but this is not a very useful feature. I recommend you do not use it.

Danger!

`C++` is a difficult language with many advanced features. Other object-oriented programming languages such as Java do not allow multiple inheritance (except of interfaces) and do not have a concept of private inheritance. This provides evidence that these features of `C++` are not widely seen to be particularly beneficial. They are best avoided.

12.6.4   Calling superclass methods

Sometimes it is useful to be able to call a superclass's methods. For example, consider an `UpAndOutOption`. If the barrier is lower than the current price, the option is worthless and so it shouldn't be priced by a slow Monte

More Sophisticated Classes 217

Carlo method. However, the rest of the time, one would want to use the default behaviour of the parent class, `KnockoutOption`, which is to price the option by Monte Carlo.

The code required to do this is as follows:

```
double price(
    const BlackScholesModel& model) const {
    if (model.stockPrice >= getBarrier())
        return 0;
    return KnockoutOption::price(model);
}
```

The key line is the last one, where we qualify the call to the function `price` with the name of the superclass `KnockoutOption`. If we didn't specify that we wanted to call the superclass's `price` function, the function would recursively call itself. The program would then crash with a stack overflow error.

12.7 Forward declarations and the structure of cpp files

The class `Shape` has a method `boundingRectangle`. This method returns a `Rectangle`. This means it is impossible to write the class `Shape` without first having told the compiler about the class `Rectangle`.

On the other hand, a `Rectangle` is a `Shape` itself, so we have a circularity. This seems to make it impossible to declare either `Rectangle` or `Shape`.

To deal with the problem you can use a *forward declaration* of the class `Rectangle`. Here's an example of how to make forward declarations:

```
class CartesianPoint;
class Rectangle;

class Shape {
public:
    /* Does the point lie in the shape */
    virtual bool contains(const CartesianPoint& point)
        const = 0;
    /* A rectangle bounding the shape */
    virtual Rectangle boundingRectangle() const = 0;
    /* By default area is computed by Monte Carlo */
    virtual double area() const;
};
```

The forward declarations:

```
class CartesianPoint;
```

218 *C++ for Financial Mathematics*

```
class Rectangle;
```

are a promise to the compiler that we will declare the classes eventually. The compiler will then allow them to be used as parameter and return types. However, it is not possible to use a class as a member variable or as a superclass until it has been fully declared.

This raises the question of what is the right order in which to put all your declarations, definitions, and so forth within a C++ file. Here are my recommendations. You should write cpp files in the following order.

1. Include statements.

2. using namespace statements.

3. Forward declarations of classes.

4. Class declarations and function declarations.

5. Function definitions and member definitions.

6. Tests (other projects often have different test conventions).

12.8 The static keyword

Sometimes you may want to associate a function with a class but not with a particular instance. Here are some reasons why you might want do this.

- The function is designed to construct instances—a so called *factory method*. We will see a detailed example of this in the next chapter.

- The function operates on vectors of instances, so you can't sensibly associate it with a particular instance.

- The function is designed to change the behaviour of every instance at once. For example, you might want to enable extra logging information for a class to help with debugging.

- The function is designed to manage all the of instances of your class. For example, since connecting to a database the first time is slow, you might want to maintain a pool of ten connections to a given database. You could then associate functions with the DatabaseConnection class that allows you to obtain and return connections from this pool.

If you want to associate a function with the class, but not with any instance, you should use the keyword static. You can also declare that a member variable is static, which means that it is shared by all instances of the class.

More Sophisticated Classes 219

As a somewhat contrived example, let us suppose that we have found a performance problem in our code. We suspect it is something to do with the evaluate function of a class we have written to compute the sin function. As part of debugging this we want to count just how often this method is called. Here is one way we could do this:

```
class CallCountedSin : public RealFunction {
public:
    static int getNumberOfCalls();
    double evaluate(double x) {
        numCalls++;
        return sin(x);
    }
private:
    static int numCalls;
};
```

Our class CallCountedSin acts like the sin function, but also counts the number of times evaluate is called on _any_ instance. We have a static member variable to save the number of calls and a static member function to read the value.

The definition of the function looks just like an ordinary member function.

```
int CallCountedSin::getNumberOfCalls() {
    return numCalls;
}
```

Notice that this function it is not allowed to access any variables other than those that are static.

Here is how we initialize the static variable numCalls:

```
int CallCountedSin::numCalls = 0;
```

This looks pretty much like a global variable definition. In a sense, that is exactly what it is. The difference is that we have associated the variable num-Calls with the class CallCountedSin. This is an improvement over having a global variable called numCalls because the static variable can be declared private. This means it is impossible for it to be used by any function in another class.

Here is an example of how you actually call a static function:

```
void testCallCountedSin() {
    CallCountedSin instance1;
    CallCountedSin instance2;
    integral(instance1, 0, 1, 1000);
    integral(instance2, 0, 1, 1000);
    int numCalls=CallCountedSin::getNumberOfCalls();
    ASSERT(numCalls == 2000);
```

220   *C++ for Financial Mathematics*

```
}
```

The point to observe is that to call the function we have to qualify its name with the name of the class.

```
int numCalls=CallCountedSin::getNumberOfCalls();
```

Since we could have many classes with identically named static methods, it makes sense that we have to qualify the name.

We have seen an example of a static variable before in Section 9.3. You may recall that we reset the seed of the random number generator using the following code:

```
mersenneTwister.seed(mt19937::default_seed);
```

This code is accessing the static variable `default_seed` that is defined in the class `mt19937`. Because the `default_seed` will be the same for all instances, it is logical to associate it with the class rather than any instance. For more complex data objects than a single integer, it may save a considerable amount of memory to have only one copy of a variable that is shared by all instances.

In summary, static functions and static variables behave exactly like global functions and global variables except for the following.

- They can be made private and have privileged access to private data. This increases encapsulation.

- They are associated with the class. Calls to static functions need to be qualified with the class name. This prevents name clashes.

Confusingly, C++ also uses the `static` keyword with functions that aren't defined in any class. In this context `static` means "cannot be used outside the current file". It is unfortunate that C++ uses the same keyword for two completely different purposes.

Danger!

Just as with other global variables, if you are using multi-threaded code then you must be careful to use locks using non-constant static variables. See Chapter 20.

12.9   The `protected` keyword

We know already that we can use the keywords `public` or `private` for information hiding. `public` means everyone can use the methods and variables. `private` means nobody can use the methods/variables including subclasses.

More Sophisticated Classes   221

Now that we are using sub-classes it is worth saying that you can also use the keyword `protected`. This means only subclasses can use the methods/-variables. In practice I recommend ignoring `protected` for the time being as it isn't particularly useful.

Exercises

12.9.1. Write `DigitalCallOption` and `DigitalPutOption` classes. Refactor `PutOption` so it extends `PathIndependentOption`.

12.9.2. Write an `AsianOption` which represents an Asian option. Show where this new class fits into the UML diagram of options.

12.9.3. Create a UML diagram for the class hierarchy with base class `Shape`.

12.9.4. There is a lot of identical code between our different chart classes. In particular they all have in common a title and two functions called `writeAs-HTML`. Design a base class and extend it for each chart. Draw a UML diagram of the resulting hierarchy. Use getter and setter methods to access the title.

12.9.5. Can you use the techniques of this chapter to improve the `Cipher` classes written for Exercise 10.6.6?

12.9.6. This exercise tests the use of static variables and functions.

(i) Write an interface class `RandomNumberGenerator` that has a function to generate a uniformly distributed random number.

(ii) Write an implementation of `RandomNumberGenerator` that uses the `mt19937` class. This design pattern of implementing an interface by simply calling another class to do all the work is called *delegation*.

(iii) Write a static function on the `RandomNumberGenerator` class called `set-Default` that sets the current default `RandomNumberGenerator` instance. Note that you should use a `shared_ptr<RandomNumberGenerator>` as the parameter to `setDefault`.

(iv) Write a static function `randUniform` on RandomNumberGenerator that allows the user to generate a uniformly distributed random number without needing to create a RandomNumberGenerator.

(v) Add a non-static `setAsDefault` function to `RandomNumberGenerator` that sets the current random number generator to be used as the default random number generator. You will need to use the `this` keyword.

(vi) In what ways, if any, do you think this improves upon the existing functions in `matlib`?

12.10 Summary

We have covered quite a few different topics in this chapter. Here is a brief review.

- Write getters and setters and use `private` data where possible.

- You can inline functions by writing the definition in the class.

- The `this` pointer makes writing setters easy. You can use it if you need a reference to the current instance.

- Build hierarchies of classes in order to inherit functionality.

- Use the `virtual` keyword to mean that a method can be overridden.

- Write =0 to mean that a function has no implementation and so create an abstract class.

- Interface classes are a special case of inheritance.

- Use forward declarations to deal with circular class declarations.

- Use `static` variables in classes instead of global variables. Use `static` functions in classes to write functions that are associated with a class in general rather than any particular instance of the class.

Chapter 13

The Portfolio Class

A key goal of this book is to show how to use C++ to price a portfolio containing a variety of different options. We will achieve that goal in this chapter.

Without object orientation, this would be very difficult to achieve. But since all our option classes have a `price` function that takes a BlackScholes-Model as a parameter it will actually be extremely easy.

We will not introduce any new language features of C++ in this chapter. The only new idea in this chapter will be a simple design pattern called the *factory pattern* that can be used to increase information hiding.

13.1 The `Priceable` interface

Financially, a portfolio is just a collection of securities such as stocks, bonds, and options in various quantities. The portfolio of an investor describes their total position across all securities.

To model a portfolio in C++ we will write a class that stores a vector of different securities and a vector of the quantities held.

We need to decide what class we will use to represent each security. This could be a stock, a bond, a derivative and so forth. The only common feature of these different securities is that they can be priced. This motivates introducing a new interface `Priceable`. Here is the interface we will use:

```
class Priceable {
public:
    /*  Compute  the  price  of  the  security  in  the
        Black--Scholes  world */
    virtual double price(
        const BlackScholesModel& model ) const = 0;
};
```

When writing a real financial system, the parameter passed to the `price` function would not be just a BlackScholesModel. In reality one would pass a more complex parameter which contains data about the entire financial market rather than data about a single stock. By using this interface for our securities, we are restricting our system to pricing securities involving only a

223

224   *C++ for Financial Mathematics*

single stock price and a fixed interest rate r. This will be sufficient to highlight the key ideas, but obviously would need to be revisited if one was writing a real trading system.

By making `ContinuousTimeOption` extend `Priceable`, we can ensure that all our options classes implement `Priceable`. We will leave writing additional classes `Stock` and `ZeroCouponBond` as exercises for the reader (see Exercise 13.5.1 for the definition of a zero-coupon bond).

13.2   The `Portfolio` interface and implementation

Now that we have designed our `Priceable` class it is easy to design the key functions of a `Portfolio`. A `Portfolio` class must have the following key functions:

(i) a function to add a `Priceable` instance together with an associated quantity;

(ii) a function to change the quantity held of a given security;

(iii) a function `price` to compute the value of the `Portfolio`.

The last item suggests that a `Portfolio` should itself implement the interface `Priceable`. This means that it will be possible to create a `Portfolio` that itself contains `Portfolios`. This accurately reflects how banks manage their books. They organise their holdings into a number of different portfolios first by line of business, then by trading desk, then by trader. Each trader will then maintain a number of separate portfolios simply as a means of organising their work.

One can see that our `Portfolio` implementation will need some way to store objects of type `Priceable` in order to keep track of the investor's position. It not possible for the `Portfolio` implementation to have member variables of type `Priceable` because `Priceable` is an abstract class. For the same reason it is impossible to have a vector of `Priceable` objects. The fundamental reason for this is that member variables and elements of vectors must all be of the same size. Different `Priceable` objects will take up different amounts of memory depending upon the variables they contain.

The solution for this is for the `Portfolio` to store data using `shared_ptr`. This unavoidable use of pointers was the reason we had to introduce the topic of pointers in Chapter 11.

This leads us to the following decisions about the design for the `Portfolio` class.

(i) Our `Portfolio` implementation will hold a vector of `shared_ptr` objects that point to `Priceable` instances.

The Portfolio Class

(ii) It will also have a vector of quantities.

(iii) Because we need to store `shared_ptr` objects, the method to add securities will take a `shared_ptr` to a security instead of a reference to the security.

Before finalising our design we add in one further consideration. The users of the `Portfolio` class don't need to know anything about how we will choose to store the data for our `Portfolio`. Therefore it would be best to keep all this inessential detail out of the header file.

With all these considerations in mind, we can now design our `Portfolio` class.

```cpp
class Portfolio : public Priceable {
public:
    /*  Virtual  destructor  */
    virtual ~Portfolio() {};
    /* Returns  the  number  of  items  in  the  portfolio*/
    virtual int size() const = 0;
    /*  Add  a  new  security  to  the  portfolio,
        returns  the  index  at  which  it  was  added  */
    virtual int add( double quantity,
            std::shared_ptr<Priceable> security ) = 0;
    /*  Update  the  quantity  at  a  given  index  */
    virtual void setQuantity( int index,
                                double quantity ) = 0;
    /*  Compute  the  current  price  */
    virtual double price(
            const BlackScholesModel& model ) const = 0;
    /*  Creates  a  Portfolio  */
    static std::shared_ptr<Portfolio> newInstance();
};
```

The `Portfolio` class is an abstract class. To create a `Portfolio` a user of this class must call the *factory method* `newInstance`. This function is guaranteed to return some kind of implementation of the `Portfolio` interface, but the user does not know precisely what class will be returned. Because we don't want the user to even know the member variables of the `Portfolio` objects we return, our factory method has to return a pointer to a `Portfolio`. As usual we use `shared_ptr` to simplify memory management.

The advantage of this is that *all* implementation details are hidden in the C++ file. Information about `private` fields and methods are not given to the user. This means we can change all of these implementation details without the user even having to recompile, they just need to link to the latest library.

The implementation is all placed in the file `Portfolio.cpp`.

```cpp
class PortfolioImpl : public Portfolio {
```

226   *C++ for Financial Mathematics*

```cpp
public:
    /* Returns the number of items in the portfolio*/
    int size() const;
    /* Add a new security to the portfolio,
       returns the index at which it was added */
    int add( double quantity,
             shared_ptr<Priceable> security );
    /* Update the quantity at a given index */
    void setQuantity( int index, double quantity );
    /* Compute the current price */
    double price(
        const BlackScholesModel& model ) const;

    vector<double> quantities;
    vector< shared_ptr<Priceable> > securities;
};
```

Placing all the implementation details in the cpp file has some other advantages. Let us describe some of them now.

(i) Our implementation class `PortfolioImpl` has a member variable with the rather complex type

```cpp
vector< shared_ptr< Priceable > >
```

This type is a vector of pointers to `Priceable` objects. It is precisely the data structure we need to store the details about the different securities that our portfolio contains. Had we declared the implementation details in a header file, the type details would be even harder to read:

```cpp
std::vector< std::shared_ptr< Priceable> >
```

This is because one should not write `using namespace` in a header file. Thus putting implementation details in a cpp file allows us to take fuller advantage of `using namespace` declarations.

(ii) In header files, the fields of all classes will normally be declared as `private` to increase encapsulation. Unfortunately, this makes testing awkward. It doesn't matter if the fields of `PortfolioImpl` are public. We've achieved perfect information hiding without the need for the `private` keyword.

(iii) The header file is much less complex. Since every file that uses `Portfolio` needs to `#include` our header file, this will reduce the compilation time of classes that use `Portfolio`. This is a small detail for this one class, but the cumulative effect across a project can be very large. The less code you put in header files, the quicker compilation will be.

The Portfolio Class

This technique of increasing encapsulation by using a factory method, `new-Instance`, to provide implementations is called the *factory design pattern*. It has numerous advantage stemming from the fact that it allows one to clearly separate the interface from the implementation. One says that this *decouples* the interface and the implementation. Dependencies between different classes in your program is called coupling. Reducing coupling results in more maintainable code.

13.2.1 Implementation of `PortfolioImpl`

All the hard work has been in the design of the classes `Portfolio` and `PortfolioImpl`. Providing the implementation of the methods is very easy.

First, we must implement the factory method `newInstance`. This simply needs to return a new `PortfolioImpl` instance.

```
shared_ptr<Portfolio> Portfolio::newInstance() {
    shared_ptr<Portfolio> ret=
        make_shared<PortfolioImpl>();
    return ret;
}
```

The implementation of `add` is easy:

```
int PortfolioImpl::add( double quantity,
            shared_ptr<Priceable> security ) {
    quantities.push_back( quantity );
    securities.push_back( security );
    return quantities.size();
}
```

as is implementing the implementation of `size`

```
int PortfolioImpl::size() const {
    return quantities.size();
}
```

and the implementation of `setQuantity`.

```
void PortfolioImpl::setQuantity( int index,
        double quantity ) {
    quantities[index] = quantity;
}
```

The end result of all our efforts is that we can write a function `price` for a Portfolio with ease:

```
double PortfolioImpl::price(
        const BlackScholesModel& model ) const {
```

228   *C++ for Financial Mathematics*

```
    double ret = 0;
    int n = size ();
    for (int i=0; i<n; i++) {
        ret += quantities[i]
                * securities[i]->price( model );
    }
    return ret;
}
```

All of this implementation code is very simple. However, pause for a moment to consider the fact that our `Portfolio` implementation can now store any kind of stock derivative and price it. It will always choose the most efficient method of pricing depending upon the type of derivative it contains. Call options will be priced using the Black–Scholes formula for call options. Put options will be priced using the put option formula. Other options will be priced by Monte Carlo. Similarly, holdings in stock and zero-coupon bonds will be priced by calling the appropriate pricing function. The end result is an extremely versatile and useful class.

What is more, our `Portfolio` will work equally well with securities we haven't even implemented yet! The user of our library can provide their own implementations of our classes if they decide to sell novel financial products. Being able to cope with rapidly changing business requirements without having to rewrite all your software is an essential feature of commercial software. It is this "pluggability" that makes object-oriented software so important in the financial industry.

13.3   Testing

As we have tried to emphasise in this book, a crucial part of software development is testing that your code works.

In this test, we check the *put-call parity* formula which we will now derive.

A portfolio, P consisting of 1 unit of a call option and -1 units of a put option both with strike K will pay off an amount equal to

$$\max\{S_T - K, 0\} - \max\{K - S_T, 0\}$$

at maturity (where S_T is the stock price at maturity). By considering separately the cases where $S_T \geq 0$ and $S_T \geq 0$, one can see that this expressions simplifies to:

$$S_T - K.$$

So the value of the P is equal to the discounted expectation of $S_T - K$. In the

Q-measure, the discounted expectation of S_T must be S_0. We deduce that the discounted expectation of $S_T - K$ is:

$$S_0 - e^{-rT} K.$$

This is the value of $\mathcal{P}$ at time 0.

The test below confirms that this holds for our Portfolio class when $r = 0$.

```
static void testPutCallParity() {
    shared_ptr<Portfolio> portfolio
        = Portfolio::newInstance();

    shared_ptr<CallOption> c
        =make_shared<CallOption>();
    c->setStrike(110);
    c->setMaturity(1.0);

    shared_ptr<PutOption> p=make_shared<PutOption>();
    p->setStrike(110);
    p->setMaturity(1.0);

    portfolio->add( 100, c );
    portfolio->add( -100, p );

    BlackScholesModel bsm;
    bsm.volatility = 0.1;
    bsm.stockPrice = 100;
    bsm.riskFreeRate = 0;

    double expected =bsm.stockPrice - c->getStrike();
    double portfolioPrice = portfolio->price( bsm );

    ASSERT_APPROX_EQUAL(100*expected,
        portfolioPrice,0.0001);
}
```

One might feel that we should write many more tests for our Portfolio class to test that it correctly prices more sophisticated portfolios. A central benefit of our design is that writing a large number of tests for the Portfolio class is not really necessary. We already have a suite of tests to confirm that various different types of option are priced correctly. The Portfolio class has a very simple price method and every line of that method is already tested by our test of put call parity. Testing that Portfolio works with a portfolio of complex options would add nothing new.

13.4 UML

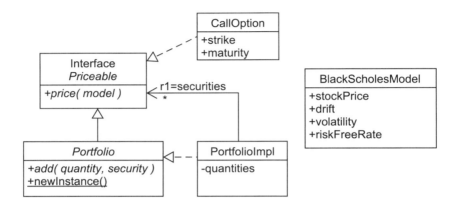

FIGURE 13.1: Design of the `Portfolio` class.

Figure 13.1 summarises the classes we have introduced in this chapter as a UML diagram.

The diagram indicates that `PortfolioImpl` implements the interface `Portfolio` which in turn extends the interface `Priceable`. `Priceable` is also implemented by the class `CallOption`.

On the other hand, a `PortfolioImpl` contains a vector of `Priceable` objects; this is indicated by the arrow labelled *securities*. This relationship is a *has a* relationship. A `PortfolioImpl` has a number of associated securities. All of these relationships are *is a* relationships and are implemented using class extension.

The * symbol on the arrow indicates that a `PortfolioImpl` can have any number of associated securities. In UML you can label the *multiplicity* of any relationship with a number or range of numbers. For example, the relationship "parents" between a human and their parents could be labelled with the multiplicity 2. * simply means there is no limit on the multiplicity.

The `newInstance` function is underlined to indicate that it is a static function.

We have deliberately omitted the full hierarchy of options from our diagram to keep it readable. We have also indicated that a CallOption has public member variables `strike` and `maturity` even though these variables can only be accessed through get and set methods. We have not mentioned that the `PortfolioImpl` stores references to `Priceable` objects using a vector

The Portfolio Class

of shared pointers. Thus our diagram is a slight simplification of our classes. It is intended to illuminate the design rather than to precisely mirror the code.

13.5   Limitations

We do not claim that the code in this chapter is the last word on pricing a portfolio of options. Here are some important limitations.

(i) We are using a fixed number of scenarios and steps in Monte Carlo calculations, this should be configurable.

(ii) We haven't made any attempt to optimise our code.

(iii) We haven't taken advantage of the multiple CPUs available on modern computers.

(iv) Our `ContinuousTimeOption` and `Priceable` interfaces are not sufficiently general. For example, they cannot cope with discrete time options and options on multiple underlying stocks.

(v) When pricing a portfolio by Monte Carlo, it isn't a good idea to keep generating random numbers. It is better to generate all the risk factors in one go and then use the same scenarios to price each element. This both involves less random number generation and will ensure that hedges cancel each other accurately.

Many of these limitations will be addressed in later chapters.

Exercises

13.5.1. A *zero-coupon bond* is a financial instrument that is guaranteed to pay out a value of 1 at its maturity T. These instruments are not often traded in practice, but are a useful mathematical abstraction to represent a risk-free bank account. Putting money in a risk-free bank account up to time T is the equivalent of investing in zero-coupon bonds with maturity T. The theory of risk-neutral pricing tells us that the risk-neutral price of a zero-coupon bond at time 0 is

$$\exp(-rT).$$

Use this fact to create implementations of the `Priceable` interface for stocks and zero-coupon bonds.

232   *C++ for Financial Mathematics*

13.5.2. Implement `UpAndInOption` and `DownAndInOption`. A portfolio containing an `UpAndInOption` and an `UpAndOutOption` should be worth the same as a portfolio containing a `CallOption`. Verify this using the `Portfolio` class. Comment on the accuracy of the answer.

13.6   Summary

We have not seen any new C++ language features while developing – `Portfolio`, just some design patterns for using C++ effectively.

- Use `shared_ptr` to build sophisticated data structures that store objects long term.

- Use the *static factory method* design pattern to maximise information hiding and reduce dependencies between files.

- Use object orientation to achieve pluggable code that will not need to be changed even when new requirements come in.

Chapter 14

Delta Hedging

In this chapter we will show how C++ can be applied to financial mathematics questions that go beyond simple risk-neutral pricing. In particular the exercises will show you how object-oriented programming techniques can be applied to solve a broad range of problems.

The problem we will address is to show numerically that the *delta hedging* strategy described in Appendix A is an effective trading strategy.

In [2], it is proved that the delta hedging strategy is completely risk free providing that various approximations to reality are used. These assumptions include that: One must be able to re-hedge one's position in continuous time; the stock price must follow geometric Brownian motion in the $\mathbb{P}$ meeasure; there must be no transaction costs; the market must be infinitely liquid; interest rates must be at a fixed level r; and so forth.

We will show in this chapter that delta hedging is still a reasonably effective strategy if one weakens the assumption of continuous-time trading to trading at fixed time points. The strategy is no longer entirely risk free, but the risk can be reduced to acceptable levels by rehedging reasonably often.

The software that we develop can then easily be enhanced to test the effects of weakening the other assumptions. In addition one can adapt the software easily to test the effectiveness of other hedging and investment strategies.

14.1 Discrete-time delta hedging

We will simulate trading in a stock that evolves according to the Black–Scholes $\mathbb{P}$-measure model. This means that we can use our existing `BlackScholesModel` and its function `generatePricePath` to simulate a stock price over a given time interval. The price of the stock at time t is denoted S_t.

At time 0, a trader sells a European call option on the stock with strike K and maturity T to a customer at the Black–Scholes price. This means that in exchange for the price P, the trader is committed to paying the customer the amount

$$\max\{S_T - K, 0\}$$

at time T.

233

234   *C++ for Financial Mathematics*

The trader's strategy is to delta hedge this liability. They delta hedge at N discrete time steps. So each time step has length $\delta t = \frac{T}{N}$.

We label the times at which the trader hedges with integers $i \in \{0, N-1\}$. So time point i takes place at time $i\delta t$. At time point i, the trader will ensure that they hold precisely Δ_i units of stock where Δ_i is the Black–Scholes delta of the call option at time point i. Any remaining money is invested in a risk-free account which grows at a fixed interest rate of r.

We write b_i for the trader's bank balance at each time point i. At time point 0 the trader puts

$$b_0 = P - \Delta_0 S_0 \tag{14.1}$$

into their risk-free account and invests the remainder of the principal, $\Delta_0 S_0$ in stock.

We can then calculate the bank balance of the trader at subsequent time points i using the difference equation

$$b_i = e^{r\delta t} b_{i-1} - (\Delta_i - \Delta_{i-1}) S_{i\delta t}. \tag{14.2}$$

This equation follows because the money they have in the bank grows at the interest rate r giving rise to the term $e^{r\delta t} b_{i-1}$ representing the amount in the bank before the trader rebalances their portfolio. The trader then needs to rebalance their portfolio to ensure they are holding Δ_i units of stock. This means that they will need to buy $(\Delta_i - \Delta_{i-1})$ units of stock. This explains the $-(\Delta_i - \Delta_{i-1}) S_{i\delta t}$ term.

At time point N, the option reaches maturity. The trader does not delta hedge. Instead they sell off any remaining stock holding and pay off the liability. This tells us that

$$b_N = e^{r\delta t} b_{N-1} + \Delta_{N-1} S_T - \max\{S - K, 0\}. \tag{14.3}$$

This final bank balance represents the profit or loss of the trader. Equations (14.1) to (14.3) have been derived using nothing more complex than the definition of the delta hedging strategy and basic accounting calculations. These calculations are no more complex than the ones you use to understand your own personal bank accounts and shopping purchases.

We can use the difference equations (14.1) to (14.3) to compute the profit and loss of the trader given a set of simulated stock prices. The one additional ingredient we need is an explicit formula for Δ_i.

By definition, $\Delta = \frac{\partial C}{\partial S}$ where C is the price of the call option and S is the current stock price. This means that one can compute Δ simply by differentiating the Black–Scholes formula (A.6) for the call price. The result is

$$\Delta = N(d_1) \tag{14.4}$$

where N and d_1 are as defined in Equation (A.6).

Together the formulae in this section allow us to explicitly calculate the profit or loss of the trader in a given scenario. By generating multiple scenarios

Delta Hedging 235

we can compute statistics for the profit and loss (PnL) of the trader. Most interestingly we can plot a histogram of the trader's PnL. We will see the C++ code required to do this in the next section.

14.2 Implementing the delta hedging strategy in C++

Given the explicit formulae in the previous section, it is straightforward to implement the delta hedging strategy. However, if one simply copies the equations into C++ one will end up with code involving variables with unhelpful names such as b that will mean nothing to someone who does not have any documentation available.

The convention in programming is to use long variable names that clearly explain what is going on. This is the opposite of the convention in mathematics of using single letter variable names.

As you will see the code below uses English names for all the variables, which makes it clear how all the terms were derived. The advantage is that this code can be understood by looking at the code alone without needing to read an explanatory document in another file.

14.2.1 Class declaration

Our first step is to declare a class that will run our hedging simulation. We will call our class HedgingSimulator. It will be a configurable class so that the user can set the value of the option to be hedged, the model to be used for the simulation, and so forth. Our design contains some flexibility in that it allows the user to set one model that should be used to generate stock prices (the simulationModel) and another that is used by the trader to compute the amount to charge the customer and the amount of stock to hold at each time (the pricingModel). By having two separate models we will be able to test what happens if the trader's beliefs about the market are incorrect.

These ideas lead us to define the following private member variables for a class called HedgingSimulator:

```
private:
    /*  The option that has been written */
    std::shared_ptr<CallOption> toHedge;
    /*  The model used to simulate stock prices */
    std::shared_ptr<BlackScholesModel>
        simulationModel;
    /* The model used to compute prices and deltas */
    std::shared_ptr<BlackScholesModel> pricingModel;
    /*  The number of steps to use */
```

236   *C++ for Financial Mathematics*

```
int nSteps;
```

The variable `nSteps` corresponds to the mathematical variable N used in the derivation of the difference equations.

We have decided to store all configuration objects using `shared_ptr`. This should be your default choice when one class needs to refer to data of another class. The alternative would be to have a member variable of type `CallOption`, but this could only store the precise class `CallOption` and not subclasses someone might write in the future. In fact it would probably be better to use a member variable of a more general type so that we can simulate hedging other types of options. See Exercise 14.4.4.

> **Tip: Reference other classes using `shared_ptr`**
>
> It is more versatile to store objects via pointer. This is because you can then hold references to subclasses if desired. This should be your default design choice when building classes that reference other classes.

Since all the member of our `HedgingSimulator` class are private, we will need set methods so that users can configure the class. We will also want a constructor to initialise the data to sensible values. Here are the associated public declarations:

```cpp
void setToHedge(
        std::shared_ptr<CallOption> toHedge) {
    this->toHedge = toHedge;
}
void setSimulationModel(
        std::shared_ptr<BlackScholesModel> model) {
    this->simulationModel = model;
}
void setPricingModel(
        std::shared_ptr<BlackScholesModel> model) {
    this->pricingModel = model;
}
void setNSteps(int nSteps) {
    this->nSteps = nSteps;
}
/* Default constructor */
HedgingSimulator();
```

We expect the user of our class to set the option to hedge, choose the models to use and the number of steps, and then ask our class to run a simulation. We provide the following public method to do this:

```cpp
/* Runs a number of simulations and returns
```

Delta Hedging 237

```
          a vector of the profit and loss */
std::vector<double> runSimulations(
         int nSimulations ) const;
```

To implement this method we will want to use three helper methods. These are private methods. The declarations are shown below:

```
    /* Run a simulation and compute
       the profit and loss */
    double runSimulation() const;
    /* How much should we charge the customer */
    double chooseCharge( double stockPrice ) const;
    /* Hoe much stock should we hold */
    double selectStockQuantity(
        double date,
        double stockPrice ) const;
```

These helper methods are introduced simply to ensure that each function is short and easy to read. For example, the function `runSimulations` simply calls the `runSimulation` multiple times and stores the results in a vector.

This completes the list of member variables and member functions of the `HedgingSimulator` class. The full class declaration can be found in the file `HedgingSimulator.h`.

14.2.2 Implementation of `runSimulation`

The most interesting code is the implementation of the function `run-Simulation`. It starts as follows:

```
double HedgingSimulator::runSimulation() const {
    double T = toHedge->getMaturity();
    double S0 = simulationModel->stockPrice;
    vector<double> pricePath =
        simulationModel->generatePricePath(T,nSteps);

    double dt = T / nSteps;
    double charge = chooseCharge(S0);
    double stockQuantity = selectStockQuantity(0,S0);
    double bankBalance = charge - stockQuantity*S0;
```

This code begins by simulating a price path in the $\mathbb{P}$-measure. It then computes the bank balance of the trader at time 0. This code is essentially equivalent to the argument used to derive Equation (14.1). The most obvious change is that we have rewritten everything with longer variable names. Another detail is that we call the function `selectStockQuantity` that decides how much stock we need to hold at each time. We also call the function `chooseCharge` that decides how much to charge the customer. By separating

238 *C++ for Financial Mathematics*

out the computation of the amount of stock required to hedge into its own function, we highlight the essential feature of the delta hedging strategy compared to other stock hedging strategies. In addition this will make it easy to enhance the code to find out what happens if the trader decides to pursue another investment strategy.

Next follows a loop where we simulate trading at each intermediate time point.

```
for (int i = 0; i< nSteps-1; i++) {
    double balanceWithInterest = bankBalance *
        exp(simulationModel->riskFreeRate*dt);
    double S = pricePath[i];
    double date = dt*(i + 1);
    double newStockQuantity =
        selectStockQuantity(date, S);
    double costs =
        (newStockQuantity - stockQuantity)*S;
    bankBalance = balanceWithInterest - costs;
    stockQuantity = newStockQuantity;
}
```

This code corresponds to Equation (14.2).

Finally we need to consider what happens at maturity. This is addressed by the following code which completes the definition of `runSimulation`.

```
    double balanceWithInterest = bankBalance *
        exp(simulationModel->riskFreeRate*dt);
    double S = pricePath[nSteps - 1];
    double stockValue = stockQuantity*S;
    double payout = toHedge->payoff(S);
    return balanceWithInterest + stockValue - payout;
}
```

This code corresponds to Equation (14.3).

14.2.3 Implementing the other methods of HedgingSimulator

The remaining details of the implementation of our `HedgingSimulator` class are extremely straightforward.

The `runSimulations` simply calls `runSimulation` multiple times and stores the results in a vector:

```
std::vector<double>
    HedgingSimulator::runSimulations(
                            int nSimulations) const {
    std::vector<double> ret(nSimulations);
    for (int i = 0; i < nSimulations; i++) {
```

Delta Hedging 239

```
        ret[i] = runSimulation();
    }
    return ret;
}
```

The default constructor ensures that a newly created HedgingSimulator has some reasonable values set for the various parameters. We choose the values so that we can create a HedgingSimulator for testing purposes without needing to do much configuration.

```
HedgingSimulator::HedgingSimulator() {
    // Choose default models and options
    shared_ptr<BlackScholesModel> model(
            new BlackScholesModel());
    model->stockPrice = 1;
    model->date = 0;
    model->riskFreeRate = 0.05;
    model->volatility = 0.2;
    model->drift = 0.10;

    shared_ptr<CallOption> option =
        make_shared<CallOption>();
    option->setStrike(model->stockPrice);
    option->setMaturity(1);

    setToHedge(option);
    setSimulationModel(model);
    setPricingModel(model);
    nSteps = 10;
}
```

The function selectStockQuantity delegates its work to a function delta on the CallOption class. This function uses Equation (14.4) to compute the delta of a call option. It is sensible to put the computation of the delta into the CallOption class as we will often want to compute the delta of an option. Ideally one should add a method to compute the delta to our entire option hierarchy (see Exercise 14.4.4).

```
double HedgingSimulator::selectStockQuantity(
        double date,
        double stockPrice) const {
    // create a copy of the pricing model
    BlackScholesModel pm = *pricingModel;
    pm.stockPrice = stockPrice;
    pm.date = date;
    return toHedge->delta(pm);
}
```

240   *C++ for Financial Mathematics*

In the code above, the line

```
BlackScholesModel pm = *pricingModel;
```

is interesting. We need to alter our pricing model so that its state reflects the state of the market at the current time step. The member variable `pricing-Model` only reflects the state of the market at the beginning of the simulation.

We do not wish to change the actual configuration of our `Hedging-Simulator`, so we do not want to change the member variable `pricing-Model`. We therefore create an entirely new variable `pm` and copy the values of `pricingModel` into that variable. We can then set the current stock price and date on the variable `pm` without changing the variable `pricingModel`. Since the member variable `pricingModel` stores the data using a pointer we have to use `*` to convert the pointer into a reference that we then copy.

The code that determines the amount to charge the customer is similar. It takes advantage of the `price` function that we have already written for call options.

```
double HedgingSimulator::chooseCharge(
                        double stockPrice) const {
    // create a copy of the pricing model
    BlackScholesModel pm = *pricingModel;
    pm.stockPrice = stockPrice;
    return toHedge->price(pm);
}
```

14.2.4   Changes to `CallOption`

We need to add a new public function to the `CallOption` class to compute the `Black--Scholes` delta. Here is the definition of that function.

```
double CallOption::delta(
    const BlackScholesModel& bsm) const {
    double S = bsm.stockPrice;
    double K = getStrike();
    double sigma = bsm.volatility;
    double r = bsm.riskFreeRate;
    double T = getMaturity() - bsm.date;

    double numerator=log(S/K)+(r+sigma*sigma*0.5)*T;
    double denominator = sigma * sqrt(T);
    double d1 = numerator / denominator;
    return normcdf(d1);
}
```

14.3 Testing the simulation

According to the theory of [2], if a trader follows the delta hedging strategy in continuous time and if the stock price follows the Black–Scholes model, then the profit and loss of the trader will be exactly zero. In practice we choose a very large value for N so that the trader is rehedging extremely frequently. We then check that the PnL of the trader is close to zero. Here is the code required to run this test.

```
static void testDeltaHedgingMeanPayoff() {
  rng("default");
  HedgingSimulator simulator;
  simulator.setNSteps(1000);
  vector<double> result=simulator.runSimulations(1);
  ASSERT_APPROX_EQUAL(result[0], 0.0, 0.01);
}
```

As usual we have seeded the random number generator by running to ensure that our test is reliable. This is the purpose of the `rng("default")`.

One can see this either as a test of our code or as numerical evidence to back up the theory of delta hedging.

14.4 Interpreting and extending our simulation

We can use our code to see how delta hedging works in practice. In particular, if a trader re-hedges infrequently, we can draw a histogram of their profit and loss. Given the library we have developed, this is easy.

```
static void testPlotDeltaHedgingHistogram() {
    rng("default");
    HedgingSimulator simulator;
    simulator.setNSteps(100);
    vector<double> result =
        simulator.runSimulations(10000);
    hist("deltaHedgingPNL.html", result, 20);
}
```

The resulting histogram is shown in Figure 14.1. This shows that although the expected profit and loss of the strategy is approximately 0, there is considerable variance. We conclude that in practice the delta hedging strategy is not risk free. This does not mean that delta hedging is not a useful strategy. It

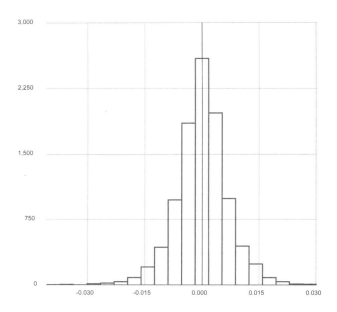

FIGURE 14.1: Histogram of the profit and loss of delta hedging, with 100 time steps and 100,000 scenarios.

simply indicates that one should charge a little more than the Black–Scholes price to compensate for the risk one needs to take in practice.

By enhancing our HedgingSimulator it is possible to see how other assumptions of the Black–Scholes model affect its conclusions. For example, what happens if one considers transaction costs? What happens if interest rates are not constant? What happens if the trader has an incorrect view about the volatility? These interesting questions are given as exercises. Some of these exercises are particularly interesting because they will require you to develop new class hierarchies to answer them.

Exercises

14.4.1. Compute the mean absolute value of the profit and loss of the delta hedging strategy. Generate a log–log plot showing the mean absolute value of the profit and loss against the number of time points at which the trader rehedges.

Delta Hedging 243

14.4.2. Why do we call `generatePricePath` rather than `generateRisk-NeutralPricePath` in our simulator?

14.4.3. What happens if there is a bid–ask spread? The ask price is the amount you must pay to purchase an asset. The bid price is the amount you receive if you sell an asset. In reality these prices are different and the difference is called the bid–ask spread.

In this question you should model the price required to buy 1 unit of stock using the process S_t as before, but you should model the price obtained if one sells 1 unit of stock as pS_t where p is some fixed proportion with $0 \le p \le 1$. How will Equation (14.2) need to be changed to account for this bid–ask spread? Change the hedging simulator so that it can be configured with a parameter p that determines the bid–ask spread. Repeat Exercise 14.4.1 but include a bid–ask spread with $p = 0.001$.

We've added a member variable `bidProportion` to `HedgingSimulator` and created a function `computePrice` which computes the money paid/money received when buying and selling stock with a fixed bid–ask spread taken into account. The simulate function then uses this `computePrice` function when calculating cash flows from buying and selling stock.

14.4.4. It would be good if `HedgingSimulator` could simulate delta hedging put options.

The first step is to add a function `delta` to `ContinuousTimeOptionBase`. There are two sensible approaches to providing a default implementation. The lazy option is to simply throw an exception saying that there is no default implementation. With more effort you could use a Monte Carlo method as described in Exercise 9.3.5. Either way, you should also provide a specialised implementation for `PutOption` which uses the formula

$$\Delta = N(d_1) - 1$$

which holds for European put options. N and d_1 are defined as in Equation (A.6).

Once you have done this, how would you change `HedgingSimulator` so you can simulate hedging put options? Write a test to check that your solution is correct.

14.4.5. What happens if the trader is wrong? The actual path followed by stock prices may not follow geometric Brownian motion at all, or it may follow geometric Brownian motion but with a different volatility.

Introduce a new interface `StockPriceModel` which can be used to generate stock prices in the $\mathbb{P}$ measure. You should ensure that `BlackScholesModel` implements the interface. Provide an alternative implementation of `Stock-PriceModel` called `TwoLevelModel` that is similar to the Black–Scholes model except that the volatility sigma takes two different values. Up to time $T/2$ the

244 *C++ for Financial Mathematics*

volatility should be equal to some value σ_1. and at subsequent times it should equal σ_2.

How would you change `HedgingSimulator` so that it can cope with different models?

14.4.6. It is interesting to consider what the profit and loss would be if one followed hedging strategies other than delta hedging. Create an interface called `Strategy` that provides functions to choose how much to charge the customer and to decide how much stock to hold at each moment in time.

You should provide four implementations of the `Strategy` interface.

(i) The class `BondsOnlyStrategy` should simulate a trader who decides to charge the Black–Scholes price and then put all the money into a risk-free account. They never rehedge.

(ii) The class `StockOnlyStrategy` should simulate a trader who decides to charge the Black–Scholes price and then put all the money into stocks. They never rehedge.

(iii) The class `DeltaHedgingStrategy` should simulate a trader who follows the delta hedging strategy.

Modify `HedgingSimulator` so you can choose your strategy. Plot histograms of the result of these hedging strategies when writing a Put Option with the following sets of parameter values:

(i) $N = 100$, $\sigma = 0.5$, $\mu = 0.01$, $T = 1$, $S_0 = 1$, $K = 1$, $r = 0$.

(ii) $N = 100$, $\sigma = 0.01$, $\mu = 0.5$, $T = 1$, $S_0 = 1$, $K = 1$, $r = 0$.

Interpret your results.

14.5 Summary

We have developed a C++ trading simulator to test the effectiveness of the delta hedging strategy. Our results provide numerical evidence that the theory of appendix A is correct.

In the exercises we showed how object-oriented programming techniques can be used to make our trading simulator extremely versatile.

Chapter 15

Debugging and Development Tools

Our software project has now become large and complex. It is still nothing like as complex as a bank's actual trading system, but it already spans thousands of lines of code and dozens of files. The need for some additional development tools to help us work productively is becoming obvious.

Because development tools are not standardised, we cannot describe in detail how to use all the different development tools you may find useful. We will focus on one standard tool called a "debugger". A debugger is an essential part of any C++ developer's toolkit. Its primary purpose is that it allows you to quickly locate where in your code an error is occurring. We will indicate how you can use the standard Visual Studio debugger on Windows and the standard gdb tool on Unix to help find bugs in your code. Other debuggers work in a broadly similar fashion, but you will need to consult their documentation for fuller details.

To show how a debugger can be used to find bugs, FMLib15 compiles but does not work correctly. We will show how to find the bugs that have been deliberately introduced.

A debugger is by no means the only development tool you should be using. We finish the chapter with a description of some other development tools you might want to use to increase your productivity. We will also discuss some important software development best practices which are crucial to working with large software projects.

15.1 Debugging strategies

Before explaining how to use a debugger, let us describe a number of strategies and techniques you can use to debug your code.

15.1.1 Unit tests

Your code should contain a lot of unit tests. Unit tests are the most valuable debugging tool you have available.

This means that whenever you introduce a bug, your tests should hopefully

245

246   *C++ for Financial Mathematics*

tell you about it quickly. This is important because it is much easier to fix a bug shortly after you have written it. You may even find that you can hardly remember precisely what code changes you have made after a couple of hours have passed. This will obviously make debugging much harder. To help with this, write lots of unit tests and make sure all your tests are run whenever you change your code (see Section 15.4.5).

As important as telling you which parts of your code are currently not working, unit tests tell you which parts of your code *are* working. If you have a million lines of code, knowing which parts of code you need to think about is extremely important.

Whenever you discover a bug in your code that *wasn't* picked up by a unit test you should write a unit test to make sure it never happens again. You should never write new code unless you are also writing tests for it.

Tip: Write a little, test a little.

Do not write large amounts of code in one sitting. Write a small amount of code. Write some tests. Make sure everything compiles and all the tests pass. Then back up your code using your version control system (see Section 15.4.1). That way when an error occurs you should have a clear idea of what you are doing that might have caused it.

15.1.2   Reading your code

Often the quickest way to find a bug in your code is to read it through. This approach will only work if you know already which function contains the error as you cannot possibly re-read all your code every time a bug occurs. However, if you are following the strategy of writing a little and testing a little, you can usually guess that the bug is in the code you have just written.

Very often it helps to explain your code to someone else. A very effective development practice is to write your code with a colleague. This is called pair programming [1]. You will find that you and your colleague both learn from each other and that many bugs are spotted before the code is even executed.

Some people suggest an alternative practice called code-review, where developers read each other's code and comment on it. This is much less enjoyable than pair programming and, it is argued in [1], much less effective.

15.1.3   Logging statements

You have probably already discovered for yourself the strategy of inserting logging statements throughout your code. By printing out the values of different variables at each line, one can often quickly work out what is going wrong.

You may find that you are regularly inserting and deleting logging state-

Debugging and Development Tools 247

ments from your code as bugs are found and fixed. The `DEBUG_PRINT` macro from our testing framework is meant to help with this as it allows you to switch debugging on and off without needing to actually delete the lines. See Section 15.4.6 for a brief discussion of other logging frameworks.

15.1.4 Using a debugger

The error messages produced by C++ when it is running can be very unhelpful. Sometimes a program will crash without providing any clue as to where in the code the problem occurred.

A *debugger* is a tool that helps alleviate this problem.

Recall that the compiler turns your C++ code into machine code. During this process, the code turns from something you wrote and understand into something that you didn't write and don't understand. This is a serious problem. To alleviate the problem, you can compile your code with additional debugging information that links your source code to the machine code that is being executed.

If you have compiled your program with the necessary debugging information, then it is possible to run your program inside a debugger. The debugger will then allow you to step through your code line by line to see what is happening. It will also translate between the machine code and your source code so you can understand what is going on. In addition you can tell the debugger to pause the execution whenever an error is detected. This allows you to gather detailed information about the cause of the error. We will describe how to use the Visual Studio debugger on Windows in Section 15.2, and how to use the GDB debugger on Unix in Section 15.3. Other debuggers work in a similar way. You should consult their documentation.

15.1.5 Divide and conquer

If you have a difficult bug to fix, it can be very useful to know the following algorithm for debugging code.

(i) By considering which of your unit tests pass and which fail, decide which area of your code contains the error.

(ii) Consider how to divide the code that contains the bug into two roughly equally sized pieces. What tests can you run to determine which of these two pieces actually contains the error?

(iii) Write the tests. Repeat this process until you have fixed the bug.

The important point is that one should not look at all your code and read through it. This approach is very time-consuming and unlikely to work. Instead, this approach uses a *divide and conquer* strategy. At each stage we halve the size of the problem.

248   *C++ for Financial Mathematics*

To see the difference, imagine your program contains one million lines of code. Reading the code through clearly takes one million steps. Let us be honest, you will fall asleep and miss the bug. The divide and conquer strategy might require writing $\log_2(1000000) \approx 20$ tests before you find the problem. This is clearly much better. It has the additional advantage that you will be increasing the size of your test suite as you work. This means that next time a bug occurs, you should be able to find it far more quickly.

15.2   Debugging with Visual Studio

This section describes how to use the debugger built into Visual Studio. If you are developing on Unix using command line tools, you should read Section 15.3 instead.

If you are using some other debugger, you should consult its documentation for the precise details of how to debug your code. Nevertheless, this section may be useful to you as an indication of the key things to look for when debugging.

When debugging your code in Visual Studio, you should select "Debug" from the drop-down menu on the Toolbar. This drop-down menu gives the choice of compiling and running either "Debug" or "Release" versions of your software. As the names suggest, the Debug version is very useful for development purposes. The Release version eliminates most of the code that helps with debugging, resulting in software that runs considerably faster.

Before launching the debugger, you should configure the debugger so that it will pause the execution of our program whenever an exception is thrown by our code. To do this select **Debug $\rightarrow$ Exceptions...** from the main menu. Then click the box to indicate that the debugger should pause the execution whenever C++ exceptions are thrown.

You can now compile and run the program in the debugger by pressing **F5**. If you compile and run FMLib in this way, an error will occur and the execution of the program will by paused. It is important to know that when you start the debugger by pressing **F5**, Visual Studio will operate as a debugger rather than as a compiler until your program has exited. This means that you shouldn't try and edit or compile your code until your program has stopped. If you want to stop debugging at any point press **Shift + F5** or select **Debug $\rightarrow$ Stop debugging** from the main menu. However, we do not want to stop debugging just yet. Instead let us read the error messages and use the debugger.

15.2.1   Obtaining a stack trace in Visual Studio

You will see an error message similar to the one shown in Figure 15.1. This error message tells you that the error was detected in the file **vector** at line

1201. The file `vector` is part of the C++ library, so we can be confident that the bug didn't actually *occur* in the `vector` class, but that it was *detected* by the `vector` class. A useful rule of thumb is that if an error occurs, it is your code that is wrong and not the libraries that you are using.

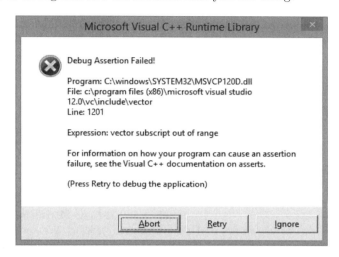

FIGURE 15.1: An assertion failed message generated on Windows.

If you click **Retry**, another window will appear asking you if you want to break or continue. "To break" simply means to pause the execution of your code so you can view it in the debugger. This is precisely what we want to do, so select **Break**.

The execution of your program is now paused and you can use Visual Studio to find out what caused the error. The most interesting part of the display is a Window labelled **Call Stack**. It should look as shown in Figure 15.2.

FIGURE 15.2: The call stack at the first bug in FMLib.

The call stack shows you what functions are executing. The second-to-last line of the call stack tells us that the function `main` is executing. It is currently on line 44 at which point it called `testHedgingSimulator`. You can see this

250   *C++ for Financial Mathematics*

because this is on the third-to-last line of the call stack. The function `test-HedgingSimulator` called `testDeltaHedgingMeanPayoff`. Continuing in this way, one can see all the functions that had yet to complete their processing when the error was detected.

The very top of the call stack is shown in grey. This is because there is no debugging information available for these lines of code. The reason for this is that the code is in a library that wasn't compiled in debug mode, specifically the `vector` library.

The most profitable way to analyse a stack trace is from the top down. You should work down the call stack until you see some code that you have written. This is then the most likely place where the bug has occurred. In this case it is line 64 of the function `runSimulation` in `HedgingSimulator`.

If you double click on the relevant line in the call stack it will bring up that line of code for you to examine.

```
double S = pricePath[nSteps];
```

The cause of the problem should be made fairly obvious by your looking at the neighbouring code. We are accessing an element of `pricePath` at index `nSteps` but in fact the price path is only `nSteps` long. We see that the code should actually read:

```
double S = pricePath[nSteps - 1];
```

To fix the problem, first stop the debugger by pressing **Shift + F5**. Now correct the code and run it once again by pressing **F5**. You will find that another error now occurs. This is because we have fixed the first bug in FMLib but there are two more to find.

15.2.2   Breakpoints and single stepping in Visual Studio

The printout that appears in the console window before the debugger pauses the program at our second bug should look similar to the following:

```
Calling testDeltaHedgingMeanPayoff()
ASSERTION FAILED
Expected -5.90148e+300
Actual 0
hedgingsimulator.cpp:106:
```

All of this printout except the last line is generated by our testing library. It tells us that an `ASSERT_APPROX_EQUAL` statement has failed at line 106 of `HedgingSimulator.cpp`.

If we view that code in our text editor we will see that a problem has arisen when running `testDeltaHedgingMeanPayoff`.

```
static void testDeltaHedgingMeanPayoff() {
    rng("default");
```

Debugging and Development Tools

251

```
HedgingSimulator simulator;
simulator.setNSteps(1000);
vector<double> result
    = simulator.runSimulations(1);
ASSERT_APPROX_EQUAL(result[0], 0.0, 0.01);
}
```

We can see that we expected the PnL of the hedging simulation to be approximately zero, but in fact it was the ridiculously large number -5.90148×10^{300}. There is clearly a mathematical error in the runSimulations function of our HedgingSimulator class. In this case, if you use the debugger to break the execution at this point and look at the call stack, you will discover that it reveals nothing we haven't been able to deduce already.

To make further progress, the first thing to do is to read through the runSimulations code to see if there is any obvious problem. Here's the code:

```
std::vector<double>
  HedgingSimulator::runSimulations(int nSimulations)
            const {
    std::vector<double> ret(nSimulations);
    for (int i = 0; i < nSimulations; i++) {
        ret[i] = runSimulation();
    }
    return ret;
}
```

There are no actual mathematical calculations in this code, so it doesn't seem likely that this is the problem. It seems a good guess that the runSimulation method is the problem. You can see the code for this method in Figure 15.4 on page 265.

You may find it difficult to spot the bug in this code even if you read through it line by line. The debugger can help you with this by allowing you to pause the execution at any point and print out the values of any variables you are interested in.

To do this you should quit the debugger. Before restarting we insert a new "breakpoint". This is simply a place where we would like the debugger to pause execution of the program so we can take a look at what is going on.

You should open the file HedgingSimulator.cpp and click in the right-hand margin of the window just next to line 39 where HedgingSimulator starts. You want to click at the point where a ball is shown in the margin in Figure 15.3. If you click in the correct location, a red ball will appear representing the breakpoint.

Once you have created a breakpoint, when you next run your code by pressing **F5**, the debugger will pause execution at the break point.

You can now follow what is happening in your code line by line by pressing **F10**. You can find this command on the menu as **Debug** $\rightarrow$ **Step Over**.

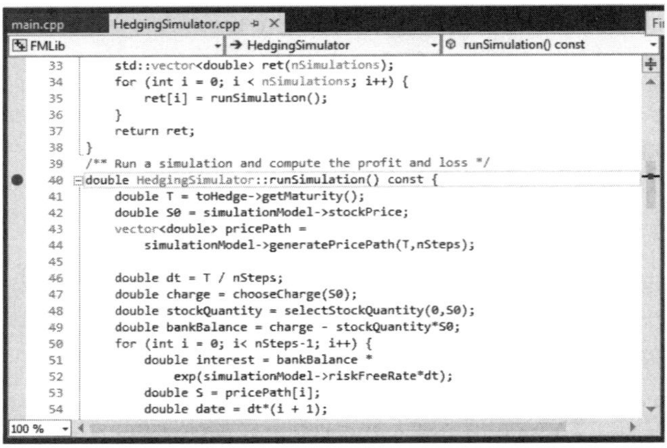

252 — C++ for Financial Mathematics

FIGURE 15.3: Inserting a breakpoint. A breakpoint is shown as a ball in the margin of your source code. You can see the ball at line 40. Click in the margin to add and remove breakpoints.

Every time you press **F10**, the next line of code will be executed. You can see the effects on the values of the local variables by looking at the area of your display labelled **Locals**.

If you step through the code line by line, you should be able to see that the calculation of interest earned is incorrect. Once you have found the problems you should stop the debugger and retest the code until `testDeltaHedging-MeanPayoff` passes.

When you have done this, there will still be one deliberate bug left in the code. Fixing this is left as an exercise.

15.3 Debugging with GDB

GDB (Gnu Debugger) is a free debugger and is the standard debugger to use on Unix systems. If you are working on Windows you should skip this section and read Section 15.4.

GDB has a command line interface which is not particularly user friendly. In practice if you are developing C++ on Unix you may want to consider using some form of "GDB front end" that makes it easier to use. For example you can integrate GDB with popular text editors such as up vim and emacs. There are also integrated development environments available, such as Eclipse. However, we will only discuss how to use GDB from the command line.

To use GDB, the first thing you must do is ensure that you have compiled your code using the debug options of `gcc`. If you are using our standard

Debugging and Development Tools 253

Makefile, then this will be done automatically. The CFLAGS variable in the Makefile uses the flags -g and -D_GLIBCXX_DEBUG for this reason. If you are not interested in debugging, but instead want your code to run as fast as possible, you should remove these options and replace them with appropriate optimisation options.

15.3.1 Using GDB to obtain a stack trace

The most important use for GDB is to find the lines of code which contain errors. FMLib15 contains some deliberate bugs. Let us see how to quickly find the cause of the first of these bugs using GDB.

Once you have compiled FMLib15, you should run it in the debugger by entering the following command in a Unix shell.

```
gdb FMLib
```

Note that we are assuming that gdb is on your path and that your current working directory contains FMLib15.

GDB will start in a state where you can enter commands to tell GDB how you want it act during this debugging session. For example, you may want to enter commands to tell it to pause the execution of your program at a particular line or to tell it to pause execution when a particular error occurs. Once you have entered these commands you will enter the command **run** and your program will be executed in the debugger.

I recommend that, at a minimum, you enter give the debugger the following instructions before executing the command **run**:

(i) Tell the debugger to pause execution if any code calls the **abort** function which terminates execution of a C++ program. Do this by typing:

```
break abort
```

(ii) Tell the debugger to pause execution if a throw statement is executed. Do this by typing:

```
catch throw
```

In addition to pausing when either of these events occurs, GDB will also pause when certain errors in your code occur. For example, if you attempt to access memory that your program doesn't have access to, a segmentation fault will occur. GDB will automatically pause when this happens without you needing to configure GDB further.

Having entered the commands above, type **run**. This will execute FMLib up to the point where the first bug is found. Execution will then be paused.

Here is the end of printout that appears when running FMLib using gdb on my computer:

```
Calling testPutCallParity()
testPutCallParity() passed.

Calling testDeltaHedgingMeanPayoff()
/usr/lib/gcc/x86_64-pc-cygwin/5.3.0/include/c++/debug/vector
    :406:error:
    attempt to subscript container with out-of-bounds index
    1000, but
    container only holds 1000 elements.

Objects involved in the operation:
sequence "this" @ 0x0xffffc7a0 {
  type = NSt7__debug6vectorIdSaIdEEE;
}

Breakpoint 1, 0x000000018012c574 in abort () from /usr/bin/
    cygwin1.dll
(gdb) backtrace
#0  0x000000018012c574 in abort () from /usr/bin/cygwin1.dll
```

The first three lines are printed out by FMLib itself. They tell us that the function `testPutCallParity` is working fine, but that a problem has occurred in the function `testDeltaHedgingMeanPayoff`. To see this simply notice that we have a logging statement saying that `testPutCallParity` has passed, but no corresponding statement for `testDeltaHedgingMeanPayoff`.

The next error tells us that the error was detected at line 406 in a C++ file called `vector`. This doesn't mean that the bug is in the file `vector`, just that it was detected there. Since `vector` is part of the standard C++ library it is very unlikely to contain any bugs. We can guess that it is our code that is using a `vector` that is at fault.

The next lines are extremely hard to interpret and not very useful for us now. Let us ignore them and move onto the last line. This tells us that the execution has stopped at `Breakpoint 1`. This means that it has stopped because of the command `break abort` that we entered earlier. Helpfully, the last line mentions it stopped in the function `abort()`. Less helpfully, the line is also telling us which library contains the `abort` function and its memory address.

To find out more about what is happening, enter the command

```
backtrace
```

This command tells the debugger to print the current state of the stack. This is simply a list of which functions are currently running. At the bottom of the stack you will see the `main` method of the program, at the top you see the current function that is executing. Here is the printout from backtrace when it was run on my computer.

Debugging and Development Tools 255

```
#0  0x000000018012c574 in abort () from /usr/bin/cygwin1.dll
#1  0x00000003f508a005 in cygstdc++-6!
    _ZNK11__gnu_debug16_Error_formatter8_M_errorEv () from /usr/
    bin/cygstdc++-6.dll
#2  0x00000001004246be in std::__debug::vector<double, std::
    allocator<double> >::operator[] (this=0xffffc7a0, __n=1000)
    at /usr/lib/gcc/x86_64-pc-cygwin/5.3.0/include/c++/debug/
    vector:406
#3  0x000000010041650f in HedgingSimulator::runSimulation (this
    =0xffffcac0)
    at HedgingSimulator.cpp:64
#4  0x0000000100416289 in HedgingSimulator::runSimulations (this
    =0xffffcac0,
    nSimulations=1) at HedgingSimulator.cpp:35
#5  0x0000000100416770 in testDeltaHedgingMeanPayoff ()
    at HedgingSimulator.cpp:105
#6  0x0000000100416bd3 in testHedgingSimulator () at
    HedgingSimulator.cpp:118
#7  0x0000000100402466 in main () at main.cpp:43
```

The bottom line tells (#7) us that the error occurred while the function `main`
was executing. The error actually occurred while line 43 of `main.cpp` was
executing. You can see this from the text at `main.cpp:43`.

Looking at the second line from the bottom, we can see that this line
must have called `testHedgingSimulator`. This in turn called `testDelta-`
`HedgingMeanPayoff`. Working through the stack in this way, you can see all
the functions that are currently executing.

Generally speaking, the bottom of the stack doesn't reveal much about
what has caused the problem. To analyse a bug, you should start at the top
and work downwards until you get to the first line of code that you have
written.

In the example above, line #0 is a call to the function abort(). The line #1
is hard to interpret, but what we can say is that it doesn't refer to any code
we've written. Line #2 is executing line 406 of `vector`. Finally on line #3 we
get to some code that we have written. It is line 64 of `HedgingSimulator.cpp`
that contains the problem.

Let us look at that line of code.

```
    double S = pricePath[nSteps];
```

The cause of the problem should be fairly obvious if you look at the neighbouring code. We are accessing an element of `pricePath` at index `nSteps`
but in fact the price path is only `nSteps` long. We see that the code should
actually read:

```
    double S = pricePath[nSteps - 1];
```

256   *C++ for Financial Mathematics*

Now that we know how to fix the bug, we should terminate the debugging session. Type `quit` to exit the debugger. You can then terminate the running process and exit the debugger.

You should now correct the code, rebuild it, and execute it again in the debugger. The code should still fail, but the stack trace should be different indicating that the bug is somewhere else.

15.3.2 Breakpoints and single stepping with GDB

The printout before the debugger pauses the program should now look as follows:

```
Calling testDeltaHedgingMeanPayoff()
ASSERTION FAILED
Expected -5.90148e+300
Actual 0
HedgingSimulator.cpp:106:
Catchpoint 2 (exception thrown), 0x00000003f512a787 in cygstdc
    ++-6!.cxa_throw
    () from /usr/bin/cygstdc++-6.dll
```

All the printout except the last line is generated by our testing library. It tells us that an `ASSERT_APPROX_EQUAL` statement has failed at line 106 of `HedgingSimulator.cpp`.

If we view that code in our text editor we will see that a problem has arisen when running `testDeltaHedgingMeanPayoff`.

```
static void testDeltaHedgingMeanPayoff() {
    rng("default");
    HedgingSimulator simulator;
    simulator.setNSteps(1000);
    vector<double> result
        = simulator.runSimulations(1);
    ASSERT_APPROX_EQUAL(result[0], 0.0, 0.01);
}
```

We can see that we expected the PnL of the hedging simulation to be approximately zero, but in fact it was the ridiculously large number -5.90148×10^{300}. There is clearly a mathematical error in the `runSimulations` function of our `HedgingSimulator` class. Unfortunately, in this case, if you run the `backtrace` command in `gdb` it will not provide much additional information.

To make further progress, the first thing to do is to read through the `runSimulations` code to see if there is any obvious problem. Here's the code:

```
std::vector<double>
  HedgingSimulator::runSimulations(int nSimulations)
                const {
```

Debugging and Development Tools 257

```cpp
        std::vector<double> ret(nSimulations);
        for (int i = 0; i < nSimulations; i++) {
            ret[i] = runSimulation();
        }
        return ret;
}
```

There are no actual mathematical calculations in this code, so it doesn't seem likely that this is the problem. It seems a good guess that the `runSimulation` method is the problem. You can see the code for this method in Figure 15.4 on page 265.

You may find it difficult to spot the bug in this code, even if you read through it line by line. The debugger can help you with this by allowing you to pause the execution at any point and print out the values of any variables you are interested in.

To do this you should quit the debugger and restart. But this time type the command

```
break HedgingSimulator::runSimulation
```

before executing the `run` command. This will then pause execution at the beginning of the `runSimulation` function.

When the program pauses you will see output similar to the following.

```
Calling testDeltaHedgingMeanPayoff()

Breakpoint 5, HedgingSimulator::runSimulation (this=0xffffc920)
    at HedgingSimulator.cpp:41
41                double T = toHedge->getMaturity();
```

This shows you that execution has paused at line 41 of `Hedging-Simulator.cpp`. The given line of code has not yet executed. To execute it, type `next`.

This should have assigned a value to the local variable T. To see what value has been assigned type `print T`.

You can now step through the function `runSimulation` line by line, checking the values of the variables as you go. You should be able to confirm that there is a problem with the computation of interest in the `runSimulation` function that you should then be able to fix.

There is still one bug remaining in the code. Fixing it is left as an exercise.

15.3.3 Other commands and features

Let us briefly mention a few useful GDB features that we have not covered.

1. The command `help` can be used to find out the names of other useful commands.

258   *C++ for Financial Mathematics*

2. The command `step` can be used to follow the execution of the code into other functions. By contrast, `next` always moves within a single function.

3. You can create a file called `.gdbinit` in your home area, which contains commands that should always be executed when gdb starts. A simple example file which would ensure gdb always pauses on calls to abort and on throw statements is shown below:

```
set breakpoint pending on
break abort
catch throw
```

Of course, GDB has many other interesting features that we have not covered. You should consult the documentation.

15.4   Other development tools and practices

A debugger is by no means the only tool you should be using to help you develop and maintain your code. In this section we will briefly review some of the important types of development tools that are available and that you should consider using.

15.4.1   Version control

It is essential to back up the source files of your project. It would be a disaster if we were to lose all the code for our project!

A source control system is used to keep a copy of all your source files to ensure that you never lose them. In addition, source control systems have some features that are essential for serious software development:

(i) A history of the changes that you have made to each file. This means that if you accidentally introduce a bug, you should always have a record of what the code looked like beforehand.

(ii) A system of labelling specific versions of your code. If you are ever going to give anyone else a copy of your software, you will want to know precisely what code they are running. By labelling each version of your source code, you can be certain what code your users are actually running.

(iii) Tools so that different developers can work on the same project and even the same files without getting in each other's way. Typically developers

Debugging and Development Tools 259

work on their own copy of the code and regularly *integrate* their changes with those of other developers. The version control system provides tools to automatically merge files together and to generate reports on the differences between different developers' copies of the code.

(iv) Tools so that it is possible to create patches to your system. A developer is usually working to add new features, but when bugs occur they need to be patched quickly without testing all the new features.

(v) Security so that only authorised users can view and edit your code.

Any serious software project will use version control. You should use some form of version control in your projects so that you at least have a full history of your changes and have regular backups.

One popular and free version-control system is called Git. It allows you to store your data on the cloud using services such as GitHub and Bitbucket. This has the advantage that issues such as backing up your data will be dealt with for you.

To their embarrassment, some major banks have forgotten to use version control for their projects! As recently as 2012 I knew of one major bank with a running system where they had lost the source code and so couldn't change or understand the system. As well as being embarrassing, this is an enormous operational risk. What would the bank do if a bug occurred?

15.4.2 Bug tracking

When bugs occur in any realistic system, they should be recorded in a bug-tracking database. As well as recording bugs, you should record requests to enhance your software.

A bug-tracking system keeps a record of what bugs are in your system, when they were fixed, who has tested them, and so forth.

On simple projects one might think that you should always fix bugs the moment that they are found, but in practice this is not always possible. For example, a bug that happens on a customer's computer may be hard to reproduce, in which case you may have little choice but to record the details and say that you will keep an eye on the problem.

For larger systems, a tool to help with recording bugs is essential. In particular it allows you to prioritise your work and focus on the most pressing problems. A bug database is also essential to keeping your customers happy. They can use it to record their complaints and monitor the progress you are making on fixing their issues.

15.4.3 Testing framework

In this book we have emphasised the importance of testing your code.

260   *C++ for Financial Mathematics*

We have introduced a simple test framework for unit testing code. In practice you might want to consider using a more elaborate testing framework. For example, you might want to use a test framework that integrates closely with your debugger so that you can jump to the right line of code whenever an error occurs. Another useful feature might be an online report showing you how many tests are being run and how many pass and fail at any moment in time.

Some test frameworks you might consider are Unit Test++, Boost library's testing framework Boost.Test, or Microsoft's Unit Testing Framework for Visual Studio.

15.4.4 Automated build

It should be possible to build, test, and deploy all your code with a single command.

In this book we have automated the process of compiling your code and running your tests for you.

If you are using Visual Studio, we have shown you how to configure your project so you can compile it and run it by pressing **CTRL+F5**. If you are working on Unix, you can build your project by typing `make` and then type the name of the project to execute it.

When working with larger projects you should make sure that you maintain a similar level of simplicity. This is not so straightforward for larger projects as they are often divided into multiple libraries. You will want to make sure that it is possible to build each library individually or to build all libraries at once. You will want to be able to build code for every operating system that your code runs on. You will want to be able to build debug versions of all your code and release versions of your code. Furthermore, some of your code may actually be automatically generated by other programs that you write.

However, it is crucial that you have a simple process for the following reasons.

(i) It should be an essentially automatic process to recreate your software from your source code. If there are written instructions that someone has to follow, there is a danger they won't be followed correctly and you will release software you cannot recreate.

(ii) It should be quick and easy to test your system. A developer new to your system should be able to build it and test it with ease. If it is hard to test your system, you won't test it much. This means it will be full of bugs.

On Windows, Visual Studio contains various tools to help automate your build. There are also numerous open source tools that can be used with any operating system such as `make`, Boost build, `cmake`, and Ant.

Debugging and Development Tools 261

15.4.5 Continuous integration

Continuous integration is a development practice where everyone working on a software project integrates their code with the work of other developers on a regular basis (often several times a day). The advantage of doing this is that any problems caused by the interactions between different people's work is caught early. It is very difficult to remember how you changed the software even a couple of hours ago, so if you don't integrate regularly it can become very challenging to merge code together.

To make continuous integration work, it is essential that you have a fully automated build that is run every time changes are made. To do this you can use a continuous integration server. A continuous integration server will:

- check your version control system to see when changes are made;

- automatically run a build;

- send emails if the build has stopped working, which say whose changes are responsible for this.

Even for a developer who is working on their own project, it is very useful to have a system that checks whether all your code is working every time you make a change. It is much better to find out what errors you have introduced at the time you introduce them.

15.4.6 Logging

Logging statements are a very useful way of understanding what is happening in your code. However, if you have too many logging statements, it becomes difficult to see the wood for the trees. Furthermore, too much logging can really slow your program down.

It is a good idea to use a lot of logging statements in your code, but to have some mechanism to increase or decrease the amount of logging that occurs as your program runs. As well as being able to record a log of what is happening as you develop code, it can be a good idea to record some sort of log of what your code is doing whenever your code runs. That way, if someone finds a bug, they can send you the log file to help you get started on debugging.

One very useful feature is to be able to click on a line in your log file and then jump to that line in your source code. Many text editors and integrated development environments can do this so long as you output your error messages using a format your environment understands. You should consult the documentation for your chosen development environment.

Tools you can use to perform logging include the Boost library's Boost.Log, and Apache's log4cxx.

15.4.7 Static analysis

There are many bugs that can be detected in `C++` code automatically.

262   *C++ for Financial Mathematics*

The build configuration that we have used performs some additional checking of our code on top of that required by C++.

In particular, we have configured our builds to fail if there is a compiler warning. We have also set the warning level so that a number of common C++ bugs will be caught automatically.

The process of looking through your code for bugs is called *static analysis*. By contrast, *dynamic analysis* would examine what happens to your code when it is actually running.

You might want to use more static analysis than we have used in this book. It is possible to ask the compiler to perform a higher level of static analysis than we have done. The danger is that the compiler will find some false positives. In other words the compiler may refuse to accept code you know does not contain any bugs. When this happens, all compilers give you a way of temporarily disabling a warning so that you can ignore these false positives. In practice it is a good idea to use more warnings and disable the warnings when necessary.

For example, writing a class without a virtual destructor is valid C++ code. For this reason compilers do not complain if you omit a virtual destructor. However, in practice your code is likely to contain memory leaks if you ever extend a class that doesn't have a virtual destructor. Therefore, if your compiler allows this, you should ask it to issue a warning whenever you extend a class without a virtual destructor. On the rare occasions when you know this isn't a problem, you can suppress the warning.

15.4.8   Memory-leak detection

Memory leaks are difficult to detect using ordinary unit tests. The reason is that memory leaks are normally only problematic once software has been running for a considerable period of time.

A good unit testing framework for C++ should be able to help with this by checking that the total memory used before and after each of your tests remains the same. In addition, there are numerous tools that you can use to help detect memory leaks in your software.

15.4.9   Profiling tools

If you have performance problems in your software then *profiling* tools can help. These allow you to find out how much time the computer spends executing different parts of your code. This then allows you to concentrate your efforts at improving performance on the areas of the code that are actually causing problems.

A mistake that many beginner programmers make is to assume that all code is performance sensitive. This is not true at all. If you use a profiler, you will discover that most (perhaps 99% or more) of the CPU effort is spent executing only a few performance-critical functions. This means that if you

Debugging and Development Tools   263

optimise the performance of almost any other function, your effort will be wasted!

Before making any serious attempt to optimise your code, it is advisable that you profile it.

Unless your performance requirements are very precise, you may not need to buy a separate piece of software to profile your code. A simple approach is to execute the slow bit of code and then stop the processing randomly ten times and record what is happening. You will probably find that the code is doing much the same thing each time you stop it. You can safely deduce that this is where the performance problem is.

It is important to say that this book is about writing good quality C++ code, but it is not about writing C++ code for absolutely optimal performance. For most problems, it is far more important that your code executes correctly than that it executes as fast as humanly possible. Most of the time it is worth sacrificing some speed for other qualities of good code such as being easy to understand, read, and extend.

Donald Knuth, a winner of the prestigious Turing Prize and a celebrated programming expert, famously said, "Premature optimisation is the root of all evil".

15.4.10 Example

This book has a complex build. The text is written using LaTeX and it refers to various C++ files in numerous different projects. Each time a change is made to this book, all of the C++ code is compiled and tested both on Windows and on Unix. In addition there are tests to check that all the code shown in the LaTeX document matches the C++ code in the computer projects. Finally, the process of putting all the code examples and text onto a website is fully automated.

Without an automated build this book would be riddled with (even more) errors. All code contains bugs, and doubtless this book contains many errors. Nevertheless, the automated build ensures that the most egregious problems are quickly eliminated.

Exercises

15.4.1. Find and fix all the bugs in FMLib15.

15.4.2. Try running FMLib without the debug information and with optimised machine code instead.

You can do this using Visual Studio by choosing the Release option from the menu bar. In addition, for experts, the project configuration options give you access to a wealth of configuration parameters, many of which can be used to tune performance.

On Unix you should remove the debugging options -g and -D_GLIBCXX_DEBUG from the Makefile and replace them with -O for optimise. You should then run `make clean`. Every time you change the options in the make file you should run `make clean`. There are various more complex options for optimisation that you can use. See the documentation of `gcc`.

How big a difference does this make to the speed at which the program runs? You might want to change the number of steps in `testPlotDeltaHedging-Histogram` to make the code slower so that you can see any difference more easily.

15.5 Summary

A debugger is a useful too for finding where the errors are in your code. Two important uses for a debugger are:

- obtaining a stack trace when an error occurs;

- stepping through code line by line to see what is going on.

While debuggers are a useful tool, they are not a substitute for unit tests. Instead they are a complimentary development technique.

There are a number of tools and techniques that you should use when writing software. These include version control and a bug database. The skills of debugging and maintaining code are at least as important as the skill of writing code in the first place. While we have not attempted to cover these issues in depth, the books [5] and [1] discuss important aspects of software development that go beyond simple coding.

Debugging and Development Tools 265

```cpp
double HedgingSimulator::runSimulation() const {
    double T = toHedge->getMaturity();
    double S0 = simulationModel->stockPrice;
    vector<double> pricePath =
        simulationModel->generatePricePath(T,nSteps);

    double dt = T / nSteps;
    double charge = chooseCharge(S0);
    double stockQuantity = selectStockQuantity(0,S0);
    double bankBalance = charge - stockQuantity*S0;
    for (int i = 0; i< nSteps-1; i++) {
        double interest = bankBalance *
            exp(simulationModel->riskFreeRate*dt);
        double S = pricePath[i];
        double date = dt*(i + 1);
        double newStockQuantity
            = selectStockQuantity(date, S);
        double costs
            = (newStockQuantity - stockQuantity)*S;
        bankBalance = bankBalance + interest - costs;
        stockQuantity = newStockQuantity;
    }
    double interest = bankBalance *
        exp(simulationModel->riskFreeRate*dt);
    double S = pricePath[nSteps];
    double stockValue = stockQuantity*S;
    double payout = toHedge->payoff(S);
    return bankBalance +interest
            +stockValue - payout;
}
```

FIGURE 15.4: Buggy code for simulating delta hedging. It may be hard to spot the errors by eye.

Chapter 16

A Matrix Class

C++ does not have a built-in class that represents a matrix of real numbers. We're going to write our own in this chapter.

During the course of writing a `Matrix` class we will cover some interesting C++ topics. Specifically,

- constructors and destructors,

- operator overloading,

- the rule of three,

- returning references,

- overloading using `const`.

16.1 Basic functionality of `Matrix`

Let us begin designing the basic behaviour of our matrix class. The `Matrix` class will store a 2-dimensional array of doubles and will have the following data members (all private).

(i) `int nrows`. The number of rows.

(ii) `int nrows`. The number of columns.

(iii) `double* data`. A pointer to the first cell.

(iv) `double* endPointer`. A pointer to one after the last cell.

The pointer data will point to a single chunk of memory of length nrows×ncols. The cell (i, j) will be stored at the location `data+(j*nRows)+i`. Note that we're using pointer arithmetic in this last expression.

We've decided to use pointers and the `new[]` operator rather than a `vector` because ultimately if you want to use special CPU optimisations (as we'll discuss in Section 16.7) you'll need to have a pointer to your data rather than just a `vector` object. In addition, it makes this `Matrix` class a good test for your pointer programming skills.

Here is the declaration of the data members of `Matrix`.

267

268   *C++ for Financial Mathematics*

```
private:

    /*   The  number  of  rows  in  the  matrix  */
    int  nrows;
    /*   The  number  of  columns  */
    int  ncols;
    /*   The  data  in  the  matrix  */
    double*  data;
    /*   Pointer  to  one  after  the  end  of  the  data  */
    double*  endPointer;
```

Our matrix class will allow the user to read the number of rows and columns but not change them. So we add the following data access methods in the header file:

```
    /*   The  number  of  rows  in  the  matrix  */
    int  nRows()  const  {
        return  nrows;
    }

    /*   The  number  of  columns  in  the  matrix  */
    int  nCols()  const  {
        return  ncols;
    }
```

These methods are inlined—that is, we write the function definition inside the class declaration. Since we'll make heavy use of the `Matrix` class it is important that we take advantage of some inlining.

We have already stated that the value of cell (i, j) is stored in the memory location `data+(j*nRows)+i`. Let's supply get and set functions so that the user can read the value of a given cell and change the value of a cell.

```
    /*   Retrieve  the  value  at  the  given  index  */
    double  get(  int  i,  int  j  )  const  {
        return  data[  offset(i,  j  )  ];
    }

    /*   Set  the  value  at  the  given  index  */
    void  set(  int  i,  int  j,  double  value  )  {
        data[  offset(i,  j  )  ]  =  value;
    }
```

These functions both take advantage of the function `offset`. This function performs the calculation of whereabouts in memory the cell (i, j) is stored relative to the point `data`.

A Matrix Class 269

```
int offset( int i, int j ) const {
    ASSERT( i >=0 && i<nrows && j>=0 && j<ncols );
    return j*nrows + i;
}
```

You will notice that this code performs an `ASSERT` statement that checks whether the coordinates are within the range of the matrix. This is great for debugging since it will find many bugs quickly. However, if you know that your code is bug free, for maximum speed you would want to ensure that this `ASSERT` statement is never called. Our testing macro `ASSERT` is designed so that it will only be called in debug mode.

16.2 The constructor and destructor of `Matrix`

The most important constructor of `Matrix` has the following declaration:

```
Matrix( int nrows, int ncols, bool zeros=1 );
```

As we'll see, it is often a waste of effort to initialise the contents of a Matrix. On the other hand, it is a common bug to fail to initialise a variable. For this reason we've allowed the user to say whether they would like our Matrix to be initialised with zeros or not. By default it is initialised with zeros. This should prevent the accidental bug, but allows the code to be optimised if desired.

Here is the definition of our constructor:

```
Matrix::Matrix( int nrows, int ncols, bool zeros )
    : nrows( nrows ), ncols( ncols ) {
    int size = nrows*ncols;
    data = new double[size];
    endPointer = data+size;
    if (zeros) {
        // memset is an optimised low level function
        // that should be faster than looping
        memset( data, 0, sizeof( double )*size );
    }
};
```

This is the first example where we've had some serious work to do in the constructor. The most important feature is that we create the in-memory storage using `new []`. We then initialise it to zero if required using `memset`.

`memset` is a standard C function which can be used to initialise a large amount of data all to the same value. Calling it may be quicker than writing a loop since CPUs often have special instructions for repetitive tasks such as this. We'll discuss this more in Section 16.7.

270   *C++ for Financial Mathematics*

Using the `new[]` operator raises a problem. When will `delete[]` be called? We need to ensure that when our `Matrix` itself is removed from memory, someone calls `delete[]`.

Our `Matrix` will be removed from memory under the following circumstances:

(i) If the `Matrix` was created by `new`, it will be removed from memory when `delete` is called.

(ii) If the `Matrix` was created by `new []`, it will be removed from memory when `delete []` is called.

(iii) If the `Matrix` was created on the stack as a local variable, it will be removed from memory when the local variable is no longer needed (i.e., when it goes out of scope).

(iv) If the `Matrix` is a member variable of another object, this will happen when the containing object is deleted.

We need a function that is called under precisely these circumstances. That is what a destructor is.

To write a destructor for your class you must follow these rules.

(i) A destructor is declared and defined just like a function except...

(ii) It must have the same name as the class except with the the addition of a tilde `~`.

(iii) It must have no return value (not even `void`).

(iv) It must have no parameters.

(v) It must not be `const`.

Here is the destructor for `Matrix`. It is inlined, so we have both declared it and defined it in the header file.

```
~Matrix () {
        delete [] data;
}
```

You don't have to inline the destructor, you can write separate declarations and definitions if preferred as

```
// declaration
class Matrix {
public:
    ...
    ~Matrix();
};
```

A Matrix Class   271

```
// definition
Matrix::~Matrix() {
    delete[] data;
}
```

You must obey certain rules when working with destructors.

Danger!

All classes that you wish to subclass should have a `virtual` destructor.[1]

Danger!

Whenever you write a destructor other than an empty `virtual` destructor you must abide by *the rule of three*. We'll cover this in Section 16.5.

You will notice that our `Matrix` class does not have a `virtual` destructor, therefore you must not subclass it. The same applies to many standard classes. For example, you should never subclass `vector<double>` no matter how tempted you may feel!

16.2.1   Virtual destructors

Every C++ class has a destructor written automatically. Irritatingly, this default destructor is not virtual. If your destructor is not `virtual` there is a danger that the wrong destructor might be called when someone attempts to `delete` a subclass.

To see how this might happen, recall that `CallOption` is a subclass of `Priceable`. This means that the following code is valid.

```
Priceable* option=new CallOption;
delete option;
```

Here we're using a pointer to a `Priceable` to delete a `CallOption`. If `Priceable` did not have a virtual destructor, C++ would not call the destructor of `CallOption` as we would like.

Although you're unlikely to write the code above, you are very likely to

---

[1]A pedant would actually say that any class that you might wish to delete using a pointer to a base class should have a virtual destructor, otherwise a virtual destructor is not necessary. If you follow the recommendation that all classes you intend to subclass should have a virtual destructor, you will follow this rule automatically.

272 *C++ for Financial Mathematics*

save a pointer to a `CallOption` in a `shared_ptr<Priceable>`. For example, our `Portfolio` class does this. The `shared_ptr` class only knows that it has a `Priceable` reference, so when it calls delete, it calls the function defined on `Priceable`.

Thus the calls to `delete` that happen due to the use of `shared_ptr` will cause problems if you create a subclass of a class that does not have a virtual destructor.

It is unfortunate that virtual destructors are not the default in C++. You won't get in much trouble if you give all classes a virtual destructor.

Having any `virtual` functions makes an object a little more complex to write in assembly language. This is because you need to carry round both the data for the object and information about its type in memory. Therefore a class without any `virtual` functions will be marginally faster than one without. So if your class isn't designed to be subclassed, it will be marginally quicker if you don't have a virtual destructor. For most classes this won't be that significant an optimisation.

Tip: Virtual destructors

If in doubt, give your class a virtual destructor.

16.2.2 When is a destructor needed?

As we've already stated repeatedly, you need to write a `virtual` destructor whenever you write a class that is designed to be subclassed.

But when do you need to write a more interesting destructor?

One situation is whenever you call `new` or `new[]` in the constructor and don't use a `shared_ptr` to store the result. More generally you should write a destructor when you obtain a resource in the constructor that you must release in the destructor.

The word resource is rather vague, but here are some examples of resources:

- a chunk of memory;

- a lock on a file that prevents others writing to the file;

- a print job that you've started;

- a connection to a database.

A resource is something that you must explicitly return to the computer when you've finished with it.

In mathematical code, using resources is quite unusual. So most of the time you won't need to write a destructors at all (except for an empty virtual one). In particular:

A Matrix Class 273

- member variables held by value are automatically deleted;

- member variables held by `shared_ptr` are automatically deleted.

This means that unless you are ignoring the advice to avoid using pointers, you probably won't have to write a destructor unless you try to interact with some code that forces you to manage resources manually. In practice this means that you will probably only have to write destructors if you wish to interact directly with C libraries.

For example, if we hadn't decided to write `Matrix` so that it used a raw pointer, we wouldn't have needed to give it a destructor.

16.2.3 Additional constructors

We've only discussed one of the constructors of `Matrix` so far. The `Matrix` class is intended to be genuinely useful, so it has a number of other constructors:

- A default constructor that creates a 1×1 matrix containing the number zero.

- A constructor that takes a `std::vector<double>` and constructs a corresponding column vector. It has an optional additional argument you can use if you want to create a row vector.

- A constructor that takes a single scalar and creates a 1×1 matrix.

- A constructor that takes a string describing the contents of the matrix.

This last constructor is interesting, not because it is challenging to implement but because it is useful design pattern to copy. Here's how you can initialise a matrix with a string:

```
Matrix m("1,2,3;4,5,6");
ASSERT( m.nRows()==2 );
ASSERT( m.nCols()==3 );
```

This makes it easy to construct new matrices to use for writing tests. It is a good idea to write your classes so that they can be initialised and configured easily for tests.

Another feature of C++ one can use to make it easier to create matrices would be to use an *initialiser list*. These will be discussed in Section 18.8.

We have not shown the code necessary for these additional interesting constructors because they do not require any new C++ language features.

While we may have many constructors, the class only has one destructor.

16.3 Const pointers

The `Matrix` class has methods `begin()` and `end()` that return pointers to the start and end of its data region.

```
/*  Access  a  pointer  to  the  first  element  */
const double* begin() const {
    return data;
}
/*  Access  a  pointer  to  the  element  after  last  */
const double* end() const {
    return endPointer;
}
/*  Access  a  pointer  to  the  first  element  */
double* begin() {
    return data;
}
/*  Access  a  pointer  to  the  element  after  last  */
double* end() {
    return endPointer;
}
```

Some of these methods return "const pointers". Recall from Chapter 11 that you can use `const` with a pointer much as you can with references. `const double* p` means that p points to a `double` but you can't change the value pointed to using p. So you can think of this as being very like a `const` reference.

Our `Matrix` function has two member functions called `begin()`. On a `const Matrix` we want `begin()` to return a `const` pointer to the data. On an ordinary `Matrix` we want `begin()` to return an ordinary pointer to the data. This will ensure that you can't get round the `const` nature of a `const` Matrix by obtaining a non `const` reference to data inside the `Matrix`.

To ensure that we have the begin function return the correct type of pointer, we need two begin functions. One is called for a `const Matrix` and the other is called for an ordinary `Matrix`.

Extra work like this is often required when working with `const`. It has to be said, this is one of those occasions where you wonder if you are working for the compiler rather than it working for you.

A Matrix Class   275

16.4 Operator overloading

16.4.1 Overloading +

Wouldn't it be wonderful if you could create two matrices and add them like this?

```
Matrix m1("1,2,3;4,5,6");
Matrix m2("2,3,4;5,6,7");

Matrix actual = m1 + m2;

Matrix expected("3,5,7;9,11,13");
expected.assertEquals( actual, 0.001 );
```

By overloading the + operator, we can make this code compile. In fact we can overload practically every C++ operator to make the matrix class much easier to work with.

(Incidentally, `assertEquals` is a helper function on `Matrix` that checks that two matrices are equal within a given tolerance.)

The main challenge in operator overloading is writing down the correct function declaration.

To overload the + operator, first notice that the + operator will need to take two parameters, both of them Matrices. More precisely, these parameters must be `const` references to Matrices.

Second, notice that the + operator will itself return a `Matrix` containing the result of the Matrix addition.

Therefore, to overload the + operator, we write a function that performs the necessary computation with the given parameters and return types. The only peculiar thing about this function is that it must have the name `operator+`.

Here is the function declaration from `Matrix.h`.

```
/*  Add two matrices
    NB - not a member function  */
Matrix operator+(const Matrix& x, const Matrix& y );
```

Notice that this is a top-level function, not a member function of Matrix. In this chapter, when we define functions for operator overloading we will clearly mark those that are top-level functions with a comment. All member functions will go uncommented.

The definition is fairly straightforward:

```
Matrix operator+(const Matrix& x, const Matrix& y ) {
    ASSERT( x.nRows()==y.nRows()
        && x.nCols()==y.nCols());
    Matrix ret(x.nRows(), x.nCols(), 0 );
```

276   *C++ for Financial Mathematics*

```
    double* dest = ret.begin();
    const double* s1 = x.begin();
    const double* s2 = y.begin();
    const double* end = x.end();
    while (s1!=end) {
        *(dest++) = *(s1++) + *(s2++);
    }
    return ret;
}
```

The only issue you are likely to have in understanding this code is in the very idiomatic use of pointers. Traditionally, C programmers write very dense code like this when working with memory. Let us try to understand it.

First, recall that *s1 means the value pointed to by s1.

Next, note that *(s1++) means compute the value pointed to by s1, *then* increment s1.

It follows that the following code:

```
*(dest++) = *(s1++) + *(s2++);
```

means add the values pointed to by s1 and s2 and store the value in dest. Next, increment dest, s1, and s2. Do you find this code hard to follow? Well, it could be written more comprehensibly as:

```
*(dest) = *(s1) + *(s2);
dest++;
s1++;
s2++;
```

But it is quite standard to write pointer code in a very dense (and unreadable?) format. Use which ever one you prefer.

One additional comment is that since we receive a const Matrix as a parameter to operator+, we must use a const double* to store the return value of begin(). We'll let all the other uses of const pointers in this chapter pass without comment from now on.

As another example, we will want to implement addition of a scalar. This adds the same scalar to every cell of the matrix. Here's the declaration:

```
/* Add a scalar to every element of a matrix
   NB - not a member function */
Matrix operator+(const Matrix& m, double scalar );
```

and here is the corresponding implementation.

```
Matrix operator+(const Matrix& m, double scalar ) {
    Matrix ret(m.nRows(), m.nCols(), 0 );
    double* dest = ret.begin();
    const double* source = m.begin();
```

A Matrix Class

```
    const double* end = m.end();
    while (source!=end) {
        *(dest++) = *(source++) + scalar;
    }
    return ret;
}
```

In fact, we want to be able to add a matrix to a scalar too, so we'll also need to declare and define this version of `operator+`

```
/*  Add a scalar to every element of a matrix
    NB - not a member function */
inline Matrix operator+(double scalar,
                              const Matrix& m ) {
    return m+scalar;
}
```

Implementing everything required for operator overloading can be time consuming, but it can result in a class that is very easy to use.

16.4.2 Overloading other arithmetic operators

We don't have to stop with overloading +, you can overload other operators too. Overloading − is much the same as overloading +. Overloading * is straightforward too apart from the fact that there are two possible choices for how to implement it.

The choice of how to overload * is a design issue and not really an issue of the C++ language. We have two choices for the meaning of *: should it mean matrix multiplication or entrywise multiplication?

Let us define these terms. Let a_{ij} be the components of an $m \times r$ matrix a and b_{jk} be the components of $r \times n$ matrix b. The standard definition of *matrix multiplication* is that the product ab is an $m \times n$ matrix with components

$$(ab)^{\text{matrix}}_{ik} = \sum_{j=1}^{n} a_{ij}b_{jk}.$$

Another possible way to define a form of multiplication for matrices is that if a_{ij} is now an $m \times n$ matrix and b_{ij} is also an $m \times n$ matrix, then the *entrywise product* is also an $m \times n$ matrix given by:

$$(ab)^{\text{entrywise}}_{ij} = a_{ij}b_{ij}.$$

This operation is also sometimes called the Hadamard product or the Schur product.

Matrix multiplication is useful if one of your matrices represents a linear transformation.

278   *C++ for Financial Mathematics*

Entrywise multiplication is useful if your matrices simply represent a collection of data points and you want to multiply the values at certain data points.

Should we use `*` to mean entrywise multiplication or matrix multiplication? There is no correct answer to this question. It is a matter of personal taste. In FMLib16, `Matrix` has a function `times` so that `a.times(b)` means entrywise multiplication. We leave `*` reserved for matrix multiplication. We haven't actually implemented matrix multiplication in FMLib16 as we don't need it yet. Implementing it is left as an exercise. The reason we don't need matrix multiplication just yet is because the data we wish to store in matrices for Monte Carlo simulations represent simple data points. As our matrices don't represent linear transformations, ordinary matrix multiplication isn't as useful as you might expect.

Tip: Use operator overloading wisely

Operator overloading can make your code easier to read if used well. Used badly, it leads to confusion.

C++ does not have a good track-record for using operator overloading wisely. For example, C++ overloads the `/` operator for integers to mean divide and round down. Most people find this confusing. Even experienced programmers can be tricked by this. Similarly C++ overloads the `^` operator to mean bitwise exclusive or when most people might expect it to mean raise to a given power. Even worse, the symbols `*` and `&` have many possible meanings in C++. These issues all add unnecessary obstacles to learning C++.

By contrast, functions with names can make their meaning immediately clear.

16.4.3 Overloading comparison operators

Overloading `>`, `>=`, `==`, `!=`, `<`, `<=` is straightforward. Here's a typical declaration. It takes two `const` references and returns a Matrix of 0's and 1's.

```
/*  Comparison  operator
    NB - not  a  member  function */
Matrix operator>(const Matrix& x, const Matrix& s );
```

Here's an example of this operator being used

```
        Matrix test1("1,2;3,4");
        Matrix test2("3,3;3,3");
        Matrix expected("0.0,0.0;1.0,1.0");
        expected.assertEquals( test1>=test2, 0.001);
```

Once again we have many various definitions for `operator>` to cope with comparing matrices to scalars as well as comparing matrices to matrices.

16.4.4 Overloading the << operator

As we know, built-in types like `double` can be written to a stream using <<. We'd like `Matrix` objects to be just as easy to print out. This can be achieved by overloading the << operator. Here is the required declaration:

```
/*  Write a matrix to a stream
    NB - not a member function */
std::ostream& operator<<(std::ostream& out,
                         const Matrix& m );
```

`operator<<` always takes an `ostream` as its first input. This is because we always have a stream on the left of <<.

The second parameter is, in this case, a `Matrix`. This is because this is the type of data we wish to print out.

The function `operator<<` returns a reference to an `ostream` that we can do some more writing to. This will in practice always be the same `ostream` that we pass in as the parameter `out`.

To see why returning the same stream is useful, consider the following code:

```
cout << "To be " << "or not to be";
```

We've written code like this several times already. This code is actually equivalent to the following:

```
(cout << "To be ") << "or not to be";
```

We have just added brackets for clarity. You can now see why the fact that `operator<<` returns a stream is useful. We're returning a stream ready to apply the << operator to again.

The implementation of `operator<<` is simple. The only point to note is that we return `out`.

```
ostream& operator<<(ostream& out, const Matrix& m ) {
    int nRow = m.nRows();
    int nCol = m.nCols();
    out <<"[";
    for (int i=0; i<nRow; i++) {
        for (int j=0; j<nCol; j++) {
            out << m(i,j);
            if (j!=nCol-1) {
                out << ",";
            }
        }
        if (i!=nRow-1) {
            out << ";";
        }
```

280   *C++ for Financial Mathematics*

```
    }
    out <<"]";
    return out;
}
```

16.4.4.1   Remarks on return by reference

We have seen that `operator<<` returns a value by a reference. We've avoided return by reference so far.

Return by reference is acceptable so long as you don't return a reference to a local variable. Return by reference saves copying, and so is more efficient than return by value. Therefore, it makes sense to use return by reference whenever it is safe to do so. Moreover, when you don't know how to copy the data (e.g., for polymorphic classes like `ostream`), return by reference is the only choice.

One effect of returning a reference is that whoever receives the reference can use that reference to modify whatever it points to. For this reason, sometimes you may want to return a `const` reference rather than just a simple reference.

As we will see in the next sub-section, sometimes we use return by reference specifically because it allows the caller to modify what the reference points to.

16.4.5   Overloading the () operator

We would like to be able to change the cells of a `Matrix` just as easily as we can change the cells of a `vector`. To achieve this we can overload the `()` operator so that you can type expressions such as `m(1,2)` to access the given row and column in matrix `m`. Here's an example of how we'd like to be able to use the `()` operator

```
    Matrix m("1,2,3;4,5,6");
    ASSERT( m(1,2)==6 ); // read a value
    m(1,2)=0; // change the value
```

When you overload the `()` operator, you must write a member function rather than a top-level function. Here are the necessary definitions to overload the `()` operator on the `Matrix` class.

```
    double& operator()(int i, int j ) {
        return data[ offset(i,j) ];
    }
```

```
    const double& operator()(int i, int j ) const {
        return data[ offset(i,j) ];
    }
```

A Matrix Class

We've written these as inlined functions for speed.

Since `operator()` returns a reference, you can change the value using the returned reference. This is what makes this line of code work:

```
m(1,2)=0; // change the value
```

Of course, you shouldn't be able to change a `const` Matrix in this way, so we need to write two versions of `operator()`: One that acts on a `Matrix` and one that acts on a `const` Matrix.

16.4.6 Overloading +=

We've already overloaded the + operator. What about the shortcut assignment operator +=? You can overload this too. Here's the necessary declaration:

```
Matrix& operator+=( const Matrix& other );
```

Just as with `operator()`, `operator+=` must be declared as a member function. When you overload assignment operators such as this, you should always return a reference to `*this`. For example, in our `Matrix` example += returns the Matrix you just called += on in just the same way as << returns the `ostream` you give it.

The reason for this convention is that it allows you to write code like this:

```
Matrix a("1,2");
Matrix b("1,2");
(a+=b)+=b;
```

You might argue that you wouldn't want to write code like this in the first place! However, it is the convention in C++ that when you overload an assignment operator, constructions like this should work. Its best to be conventional when writing code.

Here is the implementation of `operator+=`

```
Matrix& Matrix::operator+=( const Matrix& other ) {
    ASSERT( nRows()==other.nRows()
            && nCols()==other.nCols());
    double* p1=begin();
    const double* p2=other.begin();
    while (p1!=end()) {
        *p1=(*p1) + (*p2);
        p1++;
        p2++;
    }
    return *this;
}
```

The main point of interest in this code is that we return `*this` much as we return the `ostream` that is passed in when implementing <<.

16.5 The rule of three

We stated earlier in Section 16.2 that when you write a destructor, you must abide by the rule of three. We couldn't actually state the rule of three earlier because it involves overriding an assignment operator. However, we now know everything we need.

> **Tip: The rule of three**
>
> Whenever you write a destructor (other than an empty virtual destructor) you must:
>
> - override the assignment operator =;
>
> - write a copy constructor.

In fact, if you write any one of these three things:

- a non-trivial destructor;

- a copy constructor;

- an assignment operator =;

then you should write all three. While the rule of three is not a strict rule of the C++ language it is a very useful rule of thumb.

16.5.1 Overriding the assignment operator

Suppose that we have two variables of type `Matrix` called a and b. If we write the line

```
a = b;
```

we would like a and b to represent the same matrix. The operator = is called the assignment operator because it is used to assign a value to a variable.

C++ automatically gives all your classes an assignment operator. The point of the rule of three is that if your class needs a destructor then the default assignment operator written by C++ will be wrong and so must be overridden.

The default assignment operator simply copies all the variables of a into b. In the case of the matrix class, this would mean that the matrices a and b would have a member variable **data** which points to the *same* memory location. If the variable a is then deleted but not b, then the data pointed

A Matrix Class 283

to by **a** would be deleted. This would leave **b** pointing to an invalid memory address.

Here is how we override the assignment operator, `operator=` in our Matrix class.

```
Matrix& operator=( const Matrix& other ) {
    delete[] data;
    assign( other );
    return *this;
}
```

Notice the following rules for the = operator:

- The = operator should be defined as a member function.

- It should take a `const` reference and return a reference.

- You should always return `*this`.

- You should abide by the rule of three

In our example, `operator=` first deletes the data and then re-initialises all our member variables such as `nrows`, `ncols`, and `data` to match the other matrix. It does this by calling a helper function called assign.

Here is the actual implementation of `assign`.

```
void Matrix::assign( const Matrix& other ) {
    nrows = other.nrows;
    ncols = other.ncols;
    int size = nrows*ncols;
    data = new double[size];
    endPointer = data+size;
    memcpy( data, other.data, sizeof( double )*size );
}
```

Note that we don't want to set **data** to equal **other.data**. Instead we want to copy all the data from **other** so we have two duplicate matrices in memory that we can change independently.

16.5.2 Writing a copy constructor

Just as every class is automatically given an assignment operator in C++, every class is automatically given a copy constructor. This automatic copy constructor will be incorrect if your class needs a non-trivial destructor. Hence the need for the rule of three.

The copy constructor is used when you construct an object based on the values in another object of the same type. For example, the copy constructor is called at the second line in the code below:

284   *C++ for Financial Mathematics*

```
    Matrix a("1,2;3,4");
    Matrix b(a); // copy a
```

C++ will also use the copy constructor if it needs to copy data for pass by value. This means that copy constructors are actually called a lot without you noticing it.

Here is the declaration of the copy constructor for Matrix. We have chosen to inline the definition.

```
    Matrix( const Matrix& other ) {
        assign( other );
    }
```

The syntax is essentially the same as for all other constructors. Rules for a copy constructor are as follows.

- A copy constructor takes a single parameter, a **const** reference to another instance.

- It is not marked as **explicit** despite only taking one parameter.

- It performs whatever tasks are necessary to copy the data from the other reference.

In our example the **assign** function is used by both the copy constructor and the assignment operator. This is a common pattern, you'll often find that the **operator=** implementation and the copy constructor have a lot of code in common.

16.5.3   The easy way to abide by the rule of three

The copy constructor and assignment (=) operator won't be called unless either we call them explicitly or we use pass by value with instances of our class.

However, we normally pass large objects by reference, so do we really need a copy constructor and assignment operator at all? The answer, usually, is no.

Therefore, rather than writing a complex implementation of functions that will never be used, a sensible approach is to implement a **private** copy constructor and a **private** assignment operator. Rather than have the implementation of these functions do anything, they can simply **ASSERT** that they are never called. If you prefer, you can even simply fail to provide an implementation of the functions. The code will still compile and run since you actually only need to implement functions you call.

If you follow this recipe, you will have created a class that can't be copied at all. For many classes this is desirable. For example, for a class like an **ostream** implementation there should only ever be one instance writing to

A Matrix Class 285

a given file, so it doesn't make sense to allow people to create copies of an `ostream`.

Although our example of `Matrix` shows that writing a copy constructor and implementing `operator=` can be useful, in general I recommend starting with the lazy approach and only providing `public` copy methods if you have a need for them.

16.5.4 Move operators

In C++11, C++ was improved so that you can reduce the amount of copying of memory that is performed when you return data by value. This is a performance enhancement technique called `RValue` references. As a result, to write as efficient a `Matrix` implementation as possible, one should use this technique. Doing this involves writing a *move constructor* and a *move assignment* operator.

Whenever you write a class that manages its own memory, you might want to consider whether you should write a move constructor and move assignment operator for maximum efficiency. For this reason, in C++11 one can say that there is a rule of five.

In practice, as a financial mathematician, your expertise is probably in writing mathematical algorithms and not in writing data structures. For this reason we will not discuss how to implement move constructors and move assignment operators in this book.

16.6 Completing the Matrix class

We haven't listed every member function of the `Matrix` class or every function in FMLib16 for manipulating them. Here are some additional features which are implemented in FMLib16.

- Member functions `exp`, `log`, `sqrt`, `pow`, `times` to exponentiate every cell of a matrix, etc. These change the matrix itself rather than return a modified copy. This is potentially more efficient than creating a copy as less memory may be needed.

- Functions `setCol` and `setRow` to copy individual rows and columns from one matrix to another.

- Functions `row` and `col` to extract a row or column.

- Member function `positivePart` that returns $(x)^+$ for every cell x. This is handy for call options.

- `matlib` has been rewritten throughout so it works with `Matrix` rather than with `std::vector`.

- `matlib` has new functions to make it easy to work with matrices such as `ones` to create a matrices all of whose entries have the value one, `zeros` to create a matrix of zeros, `randn` to create a matrix of normally distributed random numbers, and so forth.

- Functions like `meanRows` and `meanCols` have been added to replace `mean`. It is a better design in the long-run to force the user to say whether they wish to compute the mean of the rows or columns, rather than guess based on the vector's dimensions.

If you are familiar with the GNU Octave or MATLAB you will see that the `matlib` library has been designed so it can be used in a very similar way.

A class like `Matrix` is a prime example of a highly reusable class, so it's worth going to a lot of trouble to make it as useful as possible.

16.7   Array Programming

16.7.1   Implementing an efficient matrix class

We have not attempted to write the most efficient matrix class possible. We have written the simplest C++ code that is likely to work. In practice it may be possible to write a faster matrix class, but doing so requires using techniques that are specific to the operating system you are using. Naturally enough, the fastest possible class will be carefully tuned to the hardware available.

Let us list some techniques that can be used to make a matrix class faster.

(i) *Vectorisation.* Most CPUs contain special instructions for performing repetitive tasks of the type that occur frequently in linear algebra and in Monte Carlo pricing. If our code takes advantage of these instructions it may run faster. This technique is called vectorisation. As an example the `memset` function may run faster than a simple loop that sets the values of memory directly by using these special CPU instructions. Modern compilers can vectorise a lot of your code automatically. However, for optimal performance you may wish to use either assembly language or special compiler instructions in order to take advantage of vectorisation.

(ii) *Careful memory management.* It can pay to think in detail about how memory is manipulated by a computer. The actual layout of memory on a modern computer is complex. Some data is stored on the hard drive,

A Matrix Class 287

some data is stored in read-only memory, some data is stored on the CPU either in caches or in registers. Data that is on the CPU can be accessed much faster than data stored in RAM. This in turn can be accessed much faster than data stored on the hard drive. We usually let the compiler decide on the details about how to move memory from place to place. In principle if we think about this closely, we may be able to improve performance significantly.

(iii) *Parallel processing.* A computer is able to perform multiple tasks at once. This is called parallel processing because mutliple tasks are performed in parallel. If you are using a computer with multiple CPUs, each CPU can perform calculations in parallel. So you should take advantage of this to get the maximum performance on your computer. We will discuss this topic in more detail in Chapter 20.

(iv) *GPUs.* A graphics processing unit (GPU) is a chip designed specifically for processing image data. They have been designed, for example, to perform the highly intensive computer processing required for Hollywood animations. GPUs have a very different design from traditional CPUs. GPUs are designed to perform highly repetitive tasks at speed and in parallel. CPUs are designed for general purpose computing. As a result, for the highly repetitive tasks required for linear algebra or Monte Carlo pricing, a GPU may out-perform similarly priced CPUs. GPUs have been used successfully for derivatives pricing by a number of banks.

These issues are highly specialised and the details vary considerably according to the hardware available. Moreover, as computer technology improves, the optimal way to implement your code is likely to change.

This makes it extremely challenging to write a high-performance matrix library. Fortunately it is not necessary to do so as excellent matrix libraries are available both commercially and as open-source software. We will give a brief list of possible libraries, though there are many other options you might consider.

- Eigen is an open-source library with an easy-to-use interface which has been designed to perform well across a variety of architectures.

- The Intel Math Kernel library is designed specifically for Intel processors. It is commercial software.

- cuBLAS is a linear algebra library designed for use with NVIDIA graphics cards.

16.7.2 Array programming

As is clear from the previous section, it is extremely challenging to write a high-performance matrix library. However, it is not at all difficult to use one.

288   *C++ for Financial Mathematics*

When writing any high-performance numerical library such as a Monte Carlo pricer, similar considerations will arise. This suggests that writing a high-performance Monte Carlo pricer will require a similar level of work.

The idea of *array programming* is to exploit the work others have already done writing matrix libraries in order to make your numerical code more efficient without doing too much work.

In array programming, one attempts to replace repetitive operations with matrix calculations. For example, when performing a Monte Carlo simulation with 10000 scenarios, we could replace every vector of length n with a $10000 \times n$ matrix. By writing our code so that it works with matrices representing all the scenarios at once, we can remove most of the for loops from our code and turn all the time-consuming parts of our calculation into matrix manipulations. The hope is that, if we then use a high-performance Matrix library, our Monte Carlo pricer should now perform well too.

This idea is taken to an extreme in *array programming languages* such as MATLAB. These encourage you to write every calculation using matrices. Indeeed MATLAB stands for "Matrix laboratory". Another popular example is Python with its numerical library `numpy`.

Because matrix languages encourage you to work with large matrices, they are able to perform numerical calculations extremely rapidly. There is a popular myth that C++ code is always faster than code written in other languages. In fact, numerical code written in MATLAB or Python can often be much faster than badly written C++ code. This is because of the use of the array programming technique.

It is easy to introduce array programming into FMLib, although it is admittedly a little tedious. In FMLib16 we have written new versions of all the functions in `matlib` that now operate on matrices instead of vectors. For example, we now have functions such as `sumRows` and `sumCols` where before we just had one function, `sum`.

By removing `vector`-related functionality from matlib, we force ourselves to work with the potentially more efficient `Matrix` class in all our code.

16.7.3   Array programming in the option classes

The `payoff` in the class `ContinuousTimeOption` needs to be changed to take a matrix with rows corresponding to different scenarios and columns corresponding to times. It returns a column vector with each row containing the payoff at a different time.

As an example, here is an implementation of the `payoff` function for an UpAndOutOption.

```
Matrix UpAndOutOption::payoff(
        const Matrix& prices ) const {
    Matrix max = maxOverRows( prices );
    Matrix didntHit = max < getBarrier();
```

A Matrix Class

```
Matrix p = prices.col( prices.nCols()-1);
p -= getStrike();
p.positivePart();
p.times(didntHit);
return p;
}
```

Since this method now operates on an entire Matrix of prices it will automatically benefit from any performance improvements made to the matrix library. For example, just by using a matrix library our code might now be taking advantage of GPU computing.

All of the option classes require similar vectorisation.

16.7.4 Array programming for the BlackScholesModel

```
Matrix BlackScholesModel::generatePricePaths(
        double toDate,
        int nPaths,
        int nSteps,
        double drift ) const {
    Matrix path(nPaths, nSteps,0);
    double dt = (toDate-date)/nSteps;
    double a = (drift - volatility*volatility*0.5)*dt;
    double b = volatility*sqrt(dt);
    Matrix currentLogS=log(stockPrice)*ones(nPaths,1);
    for (int i=0; i<nSteps; i++) {
        Matrix vals = randn( nPaths,1 );
        // vals contains epsilon
        vals*=b;
        vals+=a;    // vals now contains dLogS
        vals+=currentLogS; // vals now contains logS
        currentLogS = vals;
        vals.exp(); // vals now contains S
        path.setCol( i, vals, 0 );
    }
    return path;
}
```

This array programming version of `generatePricePaths` is a little harder to read than the original version.

Recall that our pricing is determined by the following difference equations:

$$s_i = s_{i-1} + \left(\mu - \frac{1}{2}\sigma^2 \right) \delta t + \sigma\sqrt{\delta t}\epsilon_i,$$

$$S_i = \exp(S_i).$$

290   *C++ for Financial Mathematics*

We can break this down into smaller pieces as follows:

$$a = \left(\mu - \frac{1}{2}\sigma^2\right)\delta t$$
$$b = \sigma\sqrt{dt}$$
$$v_1 = \epsilon_1$$
$$v_2 = b\epsilon_1$$
$$v_3 = a + b\epsilon_1$$
$$v_4 = s_{i-1} + a + b\epsilon_1 = s_i$$
$$v_5 = \exp(s_{i-1} + a + b\epsilon_1) = S_i.$$

In our code we use the same variable `vals` to store the different values v_1, v_2, v_3, v_4, and v_5. By reusing the same vector we avoid constantly allocating memory for a temporary variable which will be deleted almost immediately.

Notice the way that array programming has naturally made us reduce the number of times we compute the square root of `dt`. Previously we recomputed this for every scenario. This simple example gives a concrete hint as to why array programming might lead to faster code.

16.7.5   Array programming the Monte Carlo pricer

It is very easy to rewrite the Monte Carlo pricer to use array programming. However, there is an issue that if one wishes to use a very large number of scenarios the size of the matrices involved may become too big to comfortably fit in memory. There is a balance to be struck between the efficiency gains from array programming and the performance penalty that occurs from using too much memory.

As a result, the Monte Carlo Pricer class has been written to perform its processing in batches. Each batch uses a reasonably sized matrix of scenarios. Since the code does not require any new programming techniques we will not show it here.

FMLib16 is a version of our financial maths library that uses array programming throughout.

16.7.6   Performance

We performed some numerical experiments and found that using array programming FMLib16 actually performs a little *worse* than the original code. However, when we introduce a more complex market model with multiple stocks in Chapter 18, the array programming begins to pay off. For a multi-stock model, our array programming resulted in code that is several times faster than the naive code. These experiments were performed without using any parallel processing. Much greater improvements should be possible if one takes advantage of multiple CPUs or GPUs.

In summary, obtaining the *optimal* performance for a numerical algorithm is extremely challenging. Array programming is a way to achieve good, if perhaps suboptimal, performance with comparatively little effort.

Exercises

16.7.1. Overload the operator `<<` for all of our non-abstract option classes. This will give us helpful methods for debugging. Each option should print what type of option it is and the various parameters such as the barrier and strike. Use inheritance to avoid as much repetitive code as possible.

16.7.2. Overload the operator `()` in our RealFunction interface so that instead of writing `f.evaluate(x)` you simply write `f(x)` to evaluate a function. Which version of the class `RealFunction` do you prefer? What are the pros and cons of each version?

16.7.3. Write a class that prints to `cout`

- Whenever it is created using its default constructor;
- Whenever it has its destructor called;
- Whenever it has its assignment operator called;
- Whenever it has its copy constructor called.

Use this to check the following claims:

- The destructor is called on a local variable once the variable is no longer needed.
- The copy constructor is called when an object is passed by value.
- The copy constructor is called when an object is returned by value.
- The copy constructor is not called when an object is passed by reference.
- In the code

```
MyObject a;
MyObject b=a;
```

The copy constructor is called even though this looks like assignment.

- In this version of the code:

```
MyObject a;
MyObject b;
b=a;
```

292　　　　　　　　　*C++ for Financial Mathematics*

the default constructor then the assignment operator is called.

16.7.4.　Overload the * operator to perform matrix multiplication.

16.7.5.　Let us define some terminology. L is a pseudo-square root of a matrix, A, if $A = LL^T$.

An $n \times n$ matrix, A, is said to be positive-definite if for all non-zero vectors v, we have $v^T A v > 0$.

It turns out that all positive-definite symmetric matrices, A, have a pseudo-square root. There is a unique lower-triangular pseudo-square root called the *Cholseky decomposition* of A.

The Cholesky decomposition is useful for simulating multi-dimensional stochastic processes as described in Section 18.11. We will now describe how to compute the Cholesky decomposition mathematically. You should write the code to compute it numerically.

The matrix L can be computed using the equations:

$$L_{i,j} = \begin{cases} \sqrt{A_{j,j} - \sum_{k=1}^{j-1} L_{j,k}^2} & i = j \\ \frac{1}{L_{j,j}} \left(A_{i,j} - \sum_{k=1}^{j-1} L_{i,k} L_{j,k} \right). & i < j \\ 0 & i > j. \end{cases}$$

Although L appears on both sides of the equation, you should note that if you compute the terms of the matrix L row by row and then from left to right, you will find that each term required on the right-hand side of these equations has already been computed.

16.8　Summary

In terms of the C++ language, we have studied the following topics:

- `const` pointers.

- How to write two member functions: one that works on `const` instances; one that works on standard instances.

- How to overload operators such as `+`, `*`, and `>=`.

- How to overload the `<<` operator to make objects easy to print.

- How to overload `=`, `+=`, and `-=`. Note the special rules about what parameters these take and returning `*this`. Most C++ operators can be overloaded. For each operator there are standard best practices on how

A Matrix Class 293

it should be implemented, what the return type should be, and whether or not it should be written as a member function. Consult a text such as [13] for details.

- How to write a destructor for classes that manage memory and other resources. Note that most classes don't need a destructor.

- The rule of three: Whenever we write a destructor we write a copy constructor and override =.

Although we have implemented a sophisticated Matrix class in this chapter, we have done so purely to illustrate C++ techniques. Libraries containing much better `Matrix` implementations already exist.

Tip: Go easy on pointers and operator overloading

- Using pointers and operator overloading is justified for a `Matrix` class because we'll use it so often.

- Pointers are only really justified in our implementation for educational purposes. The code we've written would perform as well if we had used `vector`. If we had done this we would not have needed to write a destructor, copy constructor, or assignment operator.

- Operator overloading is only justified because we come with expectations on how to add and multiply matrices. Don't overload operators to give them strange meanings. This just makes code hard to read.

- Realistically, a data structure such as a `Matrix` is the kind of class you should expect to find in a library rather than spending time writing it yourself.

Chapter 17

An Overview of Templates

We have already seen data types involving angle brackets, such as

```
vector<double> v;
```

Code involving angle brackets is written using a C++ language feature called templates. In this chapter we will give a brief overview of how to write template functions and classes.

Templates are quite a difficult subject in practice. We have already said that you can think of C++ as consisting of three languages: C, an object-oriented language, and the language of templates. Learning a new language is a significant undertaking. For this reason we will only describe the basic ideas of template programming in this chapter. You should consult another reference such as [17] if you wish to write your own template library.

17.1 Template functions

Consider the following code to compute the maximum of a pair of numbers.

```
inline double findMax( double x, double y ) {
    if (x<y) {
        return x;
    } else {
        return y;
    }
}

inline float findMax( float x, float y ) {
    if (x<y) {
        return x;
    } else {
        return y;
    }
}
```

295

296   *C++ for Financial Mathematics*

This code violates the Once and Only Once principle because we have written essentially the same function twice, the only difference is that the types are different in the different function declarations.

Templates provide a solution to this problem. The following single template function allows us to find the maximum of either doubles or floats.

```
template <typename T>
inline T findMax(T x, T y) {
    if (x < y) {
        return y;
    }
    else {
        return x;
    }
}
```

When you write a templated function, you don't specify all the types of the parameters to your function. Instead, you provide a list of dummy variables, called template parameters, which can be replaced as necessary with the required types. In this example, we have one template parameter and it is called T.

In our example we specify that we are writing a template function and it has a template parameter called T, by writing:

```
template <typename T>
```

The remaining code in our template function is then a copy of the desired code to compute the maximum but using the dummy variable T to represent the type.

We can then use our `findMax` function equally well with any type that provides an implementation of <. This includes `float`, `double` but also `int` and even `string`. This is tested using the following code:

```
void testMax() {
    ASSERT( findMax(3, 1)==3); // ints
    ASSERT( findMax(2.0, 3.0) == 3.0); // doubles
    ASSERT( findMax( string("ant"),
            string("zoo")) == string("zoo"));
}
```

One unusual feature of templated functions is that the definition is normally put in the header file and not in a `cpp` file. The actual rule is that any declaration and definition should be in the same file. So if you are writing a library function, the definition must be in the header file because that is where the declaration has to be.

The reason for this rule is that the code for a template function is not fully compiled until the template is actually used. It cannot be compiled until the

An Overview of Templates 297

template is used because the types are not yet known. This means that users of your library will need to have access to the code required to define all your templated functions. This in turn means that these definitions need to be in the header file.

Danger!

If you are writing a templated library function, the definitions and declarations must be in the header files.

17.2 Template classes

The most visible use of template classes in C++ is in data types such as `vector<double>`. The syntax for template classes is very similar to that used for template functions.

Let us suppose we want to write a very simple data structure class called `SimpleVector`, which stores a vector of a fixed size n and allows the user to access the values at index i using functions `set` and `get`. We would also like to be able to use the same class to store values of any type.

To do this we write the class in the usual way, except we use a template parameter T as a placeholder for the type of data stored. Here is the code for our `SimpleVector` class.

```
template <typename T>
class SimpleVector {
public:
    /* Constructor */
    SimpleVector(int size);
    /* Destructor */
    ~SimpleVector() {
        delete[] data;
    }
    /* Access data */
    T get(int index) {
        return data[index];
    }
    /* Access data */
    void set(int index, T value) {
        data[index] = value;
    }
```

298   *C++ for Financial Mathematics*

```cpp
private:
    T* data;
    /* Rule of three - we make these private */
    SimpleVector(const SimpleVector& o);
    SimpleVector& operator=(const SimpleVector& o);
};
```

We have provided inline implementations of all methods except the constructor of our SimpleVector.

As with template functions, all definitions should be provided alongside the declarations in header files. Unfortunately you need to repeat the line declaring the template parameters you are using for every function definition. In addition you must include the template parameter in the qualifier of the name for all the functions you define. For example, here is how we have defined the constructor for SimpleVector.

```cpp
template <typename T>
SimpleVector<T>::SimpleVector( int size ) {
    data = new T[size];
}
```

As demonstrated in the unit test below, we can now use `SimpleVector` to store various different data types.

```cpp
void testSimpleVector() {
    SimpleVector<double> v1(3);
    v1.set(1, 2.0);
    ASSERT(v1.get(1) == 2.0);

    SimpleVector<int> v2(3);
    v2.set(1, 2);
    ASSERT(v2.get(1) == 2);

    SimpleVector<string> v3(3);
    v3.set(1, "Test string");
    ASSERT(v3.get(1) == "Test string");

}
```

It is possible to use multiple parameters in your templates. It is also possible to use certain simple data types such as constant integers as template parameters. One example of this is the type `array` from the library `<array>`, which is like a `vector` but of fixed length. When you create instances of `array` you must specify both the type of the data and the length. Here is an example of how you can use this type.

```cpp
void testArray() {
```

```
    array<int, 3> a;
    a[2] = 1;
    ASSERT(a[2] == 1);
}
```

Tip: You Aren't Going to Need It (YAGNI)

One of the slogans of extreme programming is "You Aren't Going to Need It". Many programmers have a tendency to write the most general software they possibly can. However, doing this can make your design more complex and harder to use. Often the extra generality goes unused. There is clearly no point writing code that will never be used.

You might be tempted to write a template version of our `matrix` class so that we can store matrices of `double`, `float` or more usefully complex numbers. However, the YAGNI slogan advises that you shouldn't bother doing this until you actually have a use for these classes.

In general, whenever you feel tempted to use templates, remember the YAGNI slogan and ask yourself if it is really going to be worth the trouble and complexity.

17.3   Templates as an alternative to interfaces

Recall that we introduced interface classes so that we could generalise our Monte Carlo pricer, so that it can work equally well with any option. We can solve this problem in a slightly different way using templates.

We can write a `monteCarloPricer` function that uses template to price options of any class using any model so long as:

(i) The option has a double valued field called `maturity`.

(ii) The option has a function `payoff` that takes the stock price at maturity as a parameter and returns the payoff of that option.

(iii) The model has a function `generateRiskNeutralPricePath` that returns a vector representing a simulated stock path in the risk-neutral measure. This function must take as parameters: the time up to which we wish to simulate the stock price path; the number of steps.

So long as all of these requirements are met, the following code will be able to price the option using the given model.

```
template <typename Option, typename Model>
double monteCarloPrice(
        const Option& option,
        const Model& model,
        int nScenarios = 10000) {
    double total = 0.0;
    for (int i = 0; i<nScenarios; i++) {
        std::vector<double> path = model.
            generateRiskNeutralPricePath(
            option.maturity,
            1);
        double stockPrice = path.back();
        double payoff = option.payoff(stockPrice);
        total += payoff;
    }
    double mean = total / nScenarios;
    double r = model.riskFreeRate;
    double T = option.maturity - model.date;
    return exp(-r*T)*mean;
}
```

This code simply *assumes* that the actual classes representing the option and the model obey all the requirements described above.

For example, the classes `CallOption` and `BlackScholesModel` meet these requirements. So this test will compile and run.

```
void testMonteCarloPricer() {
    CallOption c;
    c.strike = 110;
    c.maturity = 1;
    BlackScholesModel model;
    model.stockPrice = 100;
    model.drift = 0;
    model.riskFreeRate = 0.1;
    model.volatility = 0.2;
    model.date = 0;

    double price = monteCarloPrice(c, model);
    ASSERT_APPROX_EQUAL(price, c.price(model), 0.1);
}
```

Note that the code will still run even if `CallOption` does not implement any interfaces. What matters is the methods it has, not the interfaces it implements. For example, `BlackScholesModel` doesn't implement any particular interface, we're just assuming that the model class passed to our template function has a `generateRiskNeutralPricePath` method.

An Overview of Templates

This solution may seem simpler than the solution we gave using polymorphism. However, it suffers from a number of significant problems.

- You need to write clear documentation of how to use your template classes. In this documentation you will need to list all the functions you expect to be present in your template parameter classes and their return types. By contrast, an interface class automatically documents this information. In addition the compiler checks that the interface is correct.

- All the code must go in the header file. This makes encapsulation more challenging and increases the coupling between your files. One practical consequence of this is that your code will take longer to build.

- The compiler error messages that appear when you make a mistake with templates are very confusing. In particular the compiler will often tell you that the error is in a library file when it is in fact in your code. See Exercise 17.3.1 for an example of this.

- When you use templates, the compiler only finds the errors when the template is actually used. With object orientation, the errors are found when the code is written.

Templates can work well in mature, stable, and carefully designed and documented libraries such as the C++ standard library and Boost. They are less effective in financial software that has to be constantly adapted to changing business requirements.

Exercises

17.3.1. Change the CallOption class in our template examples library so that the word payoff is spelled incorrectly. Which file does the compiler say contains the error?

17.3.2. Write a template class `ComplexNumber` which can be used to store complex numbers of the form $x + iy$ where x and y could both be doubles, floats, or ints. The class should have a constructor which takes a single real argument and a default constructor. It should be possible to add two complex numbers with +.

17.3.3. Rewrite the integration example of Section 10.6 so that instead of using polymorphism we use a template.

17.4 Summary

Templates provide a solution to the Once and Only Once principle that is very useful for writing data structure classes. On the other hand, using templates to write unnecessarily generic code violates the YAGNI principle (You Aren't Going to Need It).

Templates can be used as an alternative to virtual functions, but virtual functions are normally the better approach.

For financial mathematics problems it is a good idea to take advantage of library classes that are built using templates, but to avoid writing your own template libraries.

Chapter 18

The Standard Template Library

The `Portfolio` class we developed in Chapter 13 can only be used to store options on a single stock. This is a significant limitation. In this chapter we would like to enhance our classes so that we can cope with a market that contains multiple stocks.

To do this we will need to use some more interesting data structures than just vectors and matrices. In particular we will want to use a data structure called a `map` which allows you to store data indexed by a key of some sort. For example, suppose you have written a class called `CompanyInfo` which contains a great deal of practical information about a company such as the address of its head office and the names of its directors. You would then want to be able to store a `CompanyInfo` object for each stock in such a way that you can quickly look up the `CompanyInfo` given just the name of the associated stock. The `map` class allows you to do this efficiently.

In Exercise 18.12.3, the same ideas will allow you to address another criticism of our `Portfolio` class. If multiple options in a `Portfolio` have the same maturity and underlying stocks, then we could use the same set of market simulations to price all of these options at once. The class `map` allows you to conveniently group data and so is a useful tool in developing more sophisticated algorithms.

C++ has a number of built-in data structures. `map` is one, as is `vector`. These standard data structures are called *containers*. The full library that includes these data structures is called the Standard Template Library (STL). It is called the Standard Template Library because all of these data structures are written using templates.

In this chapter we will look at how to use the Standard Template Library and will see how the data structures in the Standard Template Library can be used to model a market with multiple stocks.

18.1 typedef

We have already seen that the names of types in C++ can become very long. For example, we have already had to work with the following member variable of the `PortfolioImpl` class.

```
vector< shared_ptr<Priceable> > securities;
```

C++ contains a number of features that help with dealing with long type names. One is `typedef`, which we shall discuss now, a second is `auto`, which we shall discuss in the next section.

`typedef` allows you to define an abbreviation for a complex type. For example, we can define an abbreviation for a shared pointer to a `Priceable` as follows.

```
typedef std::shared_ptr<Priceable> SPPriceable;
```

The syntax looks as though we are declaring a variable called `SPPriceable` of type `shared_ptr<Priceable>` except for the keyword `typedef` in front. In other words, the syntax is:

```
typedef <<Complex Type>> <<Abbreviation>>;
```

Having defined this abbreviation, we can now rewrite the declaration of the member variable `securities` as:

```
vector<SPPriceable> securities;
```

The advantage of `typedef` in terms of reducing the amount of typing is clear. By making the code shorter, it also makes it more readable. However, excessive use of `typedef` will make your code harder to understand and debug. Each use of `typedef` introduces something extra for the reader of your code to remember.

Personally I find that the convention of having type names beginning `SP` to abbreviate `shared_ptr` is both convenient and easy to remember. I also use a prefix of `SPC` to mean a `const` pointer such as:

```
typedef std::shared_ptr<const Priceable> SPCPriceable;
```

We will use these conventions without comment from now on, although they are by no means a standard convention in C++.

Unless you have such clear conventions, it is best to avoid putting `typedef` declarations at the top level in a header file. This is because it is often more confusing than helpful. Within a `cpp` file you can allow yourself more freedom because you only need to remember the abbrevation you are using within the file.

The Standard Template Library 305

One alternative is to put the `typedef` inside a class declaration. By associating the `typedef` clearly with another class, one can often make its meaning clear. When you do this you are creating what is called a *member type*. For example, we can add a member type `sp` to the class `Priceable` to make it easier to refer to shared pointers to `Priceable`. We can then use this to add a member type `spVec`, which makes it easier to refer to vectors of shared pointers to `Priceable`. We would then be able to use the abbreviation `Pricable::spVec` to refer to this type in other classes, as shown below.

```cpp
class Priceable {
public:
    typedef shared_ptr<Priceable> sp;
    typedef vector< sp > spVec;
    // ... more code ...
};

class Portfolio {
private:
    Priceable::spVec securities;
    // ... more code ...
};
```

There is an alternative notation which is sometimes necessary to help the compiler work out whether a term such as `Priceable::spVec` refers to a static member variable or a member type. The alternative notation is simply to add the word `typename` at the front.

```cpp
    typename Priceable::spVec securities;
```

When working with templates, it is good practice to include the `typename` to ensure your code can be compiled unambiguously.

There is not much practical difference between the convention of having type names such as `SPPriceable` and the convention of having a member type called `sp`. You should follow the conventions of the project you are working on.

Member types are used heavily in the standard template library. This is not because they provide convenient abbreviations, but because they allow you to write template functions more easily.

Suppose that you want to write a templated `sumVector` function that takes a vector of numbers of some sort and computes the sum. We want our function to be able to add up the values in a vector of integers, a vector of real numbers, a vector of complex numbers, or even a vector of matrices.

Our templated function will have one template parameter `V` which will be some kind of vector class. Our function `sumVector` will take a reference to a vector of type `V` as its parameter. But what will be its return type? The return type will depend upon the type of `V`. For a complex vector the return type should be a complex number, for a real vector it should be a real number. How can we deduce the return type if all we know is V?

The solution is that vectors in C++ have a member type `value_type` which contains just the information we need. Therefore the return type of our `sumVector` function should be `V::value_type`. Because of possible ambiguities that could cause problems for the compiler we must write this out in full as `typename V::value_type` when using templates.

In full the code needed for our `sumVector` function is:

```
template <typename V>
typename V::value_type sumVector(const V& vector) {
    typename V::value_type total = 0;
    for (int i = 0; i < (int)vector.size(); i++) {
        total += vector[i];
    }
    return total;
}
```

18.2 auto

The `auto` keyword allows you to avoid typing the full name of a type when it can be deduced automatically. For example, if you want to store the return value of a function in a variable of the same type, you can use `auto`. In the code below, the compiler can work out that s must be of type `double` so we can use `auto`.

```
double d = 4.0;
auto s = sqrt(d);
```

You can use `auto` in combination with `&` and `const` to indicate whether you want a copy of the data or the reference and whether you wish to be able to modify the data. Here are some examples:

```
vector<double> vec(10,0.0);
auto& dRef = vec[5];
dRef = -1.0;
ASSERT(vec[5] == -1.0);

auto d = vec[6];
d = -1.0;
ASSERT(vec[6] == 0.0);

const auto& dRef2 = vec[7];
```

The Standard Template Library 307

```
ASSERT(dRef2 == 0.0);
```

In this example we take advantage of the fact that the [] operator on vector actually returns a reference to the cell in the vector. So dRef contains a reference and can be used to modify the vector. On the other hand, d contains a copy of the data in the vector and so cannot be used to modify the vector. Finally dRef2 is a reference but is **const** so it can only be used for read access to the vector.

These examples are rather contrived since **auto** is only marginally quicker to type than **double**. However, as we will see below, the **auto** keyword can save a great deal of typing and mental energy. You should use it extensively in your code.

18.3 Using iterators with vectors

So far in this book we have looped through vectors using a **for** loop to increment an integer index i.

There is another way of looping through the elements of a vector which works equally well for all data storage classes in the Standard Template Library. This approach is to use an *iterator*.

When you call **begin()** on a vector it returns you an object called an iterator. Iterators are classes that behave a lot like pointers in that you work with them by using operators such as *, ++ and ==. Here is an example of how you could use an iterator to sum all the elements of a vector.

```
vector<double> v({ 1.0, 2.0, 3.0 });

double sum = 0.0;
vector<double>::iterator i = v.begin();
while (i != v.end()) {
    sum += *i;
    i++;
}

ASSERT(sum == 6.0);
```

The code *looks* as though **begin** returns a pointer to the first element of the vector and **end** returns a pointer to one after the last element of the vector. We access the data referenced by the iterator using * just as we do with a pointer. We increment the iterator using ++ just as we do with a pointer. We decide when to terminate the loop by comparing the pointer with **end** just as we did when working with pointers in Chapter 16. The only

308   *C++ for Financial Mathematics*

clue that we are not actually working with a pointer is the type declaration
`vector<double>::iterator`.

The advantage of using iterators is that they can be used with all the
classes in the Standard Template Library. This means that code you write for
one container can be used with other containers. This makes it possible to
write generic algorithms using templates that can work with any container.

You can modify data using the `iterator` on a vector as shown in the
example below. This function uses an iterator to set every element of a vector
to zero.

```cpp
void setZero(vector<double>& v) {
    vector<double>::iterator i = v.begin();
    while (i != v.end()) {
        *i=0;
        i++;
    }
}
```

If you only want to view the data, you can use a `const_iterator` instead as
shown in the example below.

```cpp
double sumVector( const vector<double>& v ) {
    double sum = 0.0;
    vector<double>::const_iterator i = v.begin();
    while (i != v.end()) {
        sum += *i;
        i++;
    }
    return sum;
}
```

If you call `begin` on a `const` reference, it will return a `const_iterator`.
So in the code above it is essential that the type of i is specified as a
`const_iterator`, otherwise the code would not compile.

The distinction between an `iterator` and a `const_iterator` is a little
confusing and certainly very tedious to remember. This is where `auto` comes
into its own. The code below is equivalent to the last example, but is both
easier to understand and quicker to type.

```cpp
double sumWithAuto(const vector<double>& v) {
    double sum = 0.0;
    auto i = v.begin();
    while (i != v.end()) {
        sum += *i;
        i++;
    }
    return sum;
```

The Standard Template Library

```
}
```

18.4 `for` loops and containers

A container is just a class which stores data and follows the conventions of the standard library. At a minimum it should provide methods `begin` and `end`, which return iterators just as `vector` does.

This means that we have already written a container! The class `Matrix` from Chapter 16 has functions `begin` and `end` which return pointers. Pointers have definitions for the operators `++`, `==`, and `*` and so forth, so they count as iterators.

C++ contains a special syntax for looping through the elements of a container. For example, the following code sums all the elements of a `Matrix`.

```
Matrix matrix("1,3;2,4");
double total = 0.0;
for (auto d : matrix) {
    total += d;
}
ASSERT_APPROX_EQUAL(total, 10.0, 0.001);
```

This for-loop syntax means that for every element in the matrix, we will assign the value to d and then perform the body of the loop. This syntax is often much more convenient than writing the loop using iterators, although the end result is essentially identical.

Having `begin` and `end` functions is just the minimum requirement for a container. One should follow all of the conventions used in the standard template library which make sense for your container.

Therefore we have added some member types to our `Matrix` class that have precisely the same names as the corresponding members of `vector`.

```
typedef double value_type;
typedef double* iterator;
typedef const double* const_iterator;
```

If we follow the same naming conventions for our container class, it is possible to write functions that can work equally well with vectors and with our `Matrix` class. For example, here is a templated function that can be used to compute the sum of the elements in a `vector` or in a `Matrix`.

```
template <typename C>
typename C::value_type sumContainer(const C& c) {
    typename C::value_type total = 0;
```

```
    for (auto v : c) {
        total += v;
    }
    return total;
}
```

18.5   The container set

One useful container class is `set`. To use it you must `#include <set>`.

One key difference between a `set` and a `vector` is that a `set` does not contain duplicate elements (just like the mathematical concept of a set). Another difference is that the data stored in a `set` is always sorted.

For this reason, you cannot store arbitrary data in a `set`, you must be able to compare two elements using the operator `<`. Two elements a and b are considered to be the same if neither a<b nor b<a.

If you want to store a particular data type in a `set` you will need to overload the `<` operator. Actually, this isn't quite true. You can choose to tell the `set` to use another function other than `<` to perform the comparisons if you like, but we will only discuss the standard case where you overload `<`.

Fortunately the data types one most often wants to store in a `set`, namely numeric types and `string`, already have appropriate implementations of `<`.

Here is an example of how you can insert elements into a set using `insert`, confirm that the set does not contain duplicates, and then print out the contents of the set using the `for` syntax.

```
    set<int> ints;
    ints.insert(1);
    ints.insert(3);
    ints.insert(2);
    ints.insert(3); // duplicate ignored
    ASSERT(ints.size() == 3);
    for (auto i : ints) {
        std::cout << "Item " << i <<"\n";
    }
```

Note that if you do not mind whether the elements in the `set` are sorted, you can use an `unodered_set` instead. It is usually faster to use an `unordered_set`. See Section 18.9.2 for more information on `unordered_set`.

18.6 The container vector

We are familiar with the container vector already, but we have not given a great deal of detail about when it should be used and when other containers should be used.

A vector stores its data in memory in a single contiguous region just as happens if you create a block of memory using new []. vector improves upon creating memory using new [] yourself in the following ways.

(i) A vector stores the size of the memory block as well as just a pointer.

(ii) When you add elements to a vector, it will decide if it needs to allocate more memory and, if required, it will allocate a new block of memory and copy the data in. A vector has a *capacity* which is the size of the memory it has allocated and a *size* which is the number of elements it contains. The vector will always make sure the capacity is larger than the size by allocating new memory as needed.

(iii) In debug mode, the iterator and the operator [] contain extra checks that you are only accessing data that lies inside the vector. This is very helpful for debugging.

(iv) A vector obeys all the conventions of the standard template library so it can be used more easily than a raw pointer.

As we have seen already, vectors are easy to use. They are also very fast because they use contiguous blocks of memory which the computer finds easy to work with. In particular you can access the i-th element of a vector extremely rapidly. It does not take significantly longer than accessing the first element.

The downside of a vector is that it is very slow to insert a new item in the middle of a vector. The worst case is if you wish to insert a single item at the beginning of a vector. You will then need to copy all the old contents of the vector into new memory locations before inserting the new element. This is illustrated in Figure 18.1.

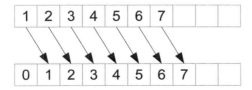

FIGURE 18.1: Inserting into a vector is $O(N)$. Blank squares indicate unused capacity.

312   *C++ for Financial Mathematics*

This means that a `vector` is not the right data type to use to store the characters in a spreadsheet document. Nobody wants it to take a long time to insert a single character at the beginning of a document.

We say that the time taken to insert an element into a vector is $O(N)$ where N is the length of the vector. On the other hand, reading a single element at a specified index is $O(1)$. In other words, reading a single item in a vector is independent of the length of the vector. Adding elements to the end of a vector is typically of $O(1)$ unless you are unlucky and the vector has to expand its capacity, in which case it is $O(N)$ as the entire vector has to be copied.

A `set` is faster than a `vector` for inserting data. Inserting is $O(\log N)$. We will explain why when we discuss the very similar `map` container in Section 18.9.1.

18.7 The container `list`

A `list` is very similar to a `vector` but it is designed so that it is easy to insert data at arbitrary points in $O(1)$. In particular, you can add data efficiently to either the front or end of a list. By contrast, with vectors you can only add data at the end efficiently.

This means that the `list` class is very useful for organising priority lists. You can put high-priority items at the front of the list and low-priority items at the end.

Under the covers, each element in a `list` is stored using a data type that looks something like this:

```
class Link {
    Data d;
    Link* next;
    Link* previous;
};
```

Each `Link` is stored in an arbitrary location in memory, but by following the pointer `next` you can always quickly find the next item in the list. If you wish to go backwards you can follow the pointer `prev`. The actual `list` class itself needs to store a pointer to the first and last elements. The situation is depicted in Figure 18.2.

This data structure is called a doubly linked list. You can quickly insert a new element by creating a new `Link` and re-routing a few of the pointers. The changes that need to be made are indicated in Figure 18.3 using bold and dashed arrows. The important point is that the number of changes is independent of the length of the list. This means that insertion takes $O(1)$.

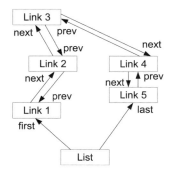

FIGURE 18.2: A doubly linked list storing the numbers 1 to 5.

However, if you want to find the i-th element of the list, you have no choice but to run through each element in sequence. This takes $O(i)$.

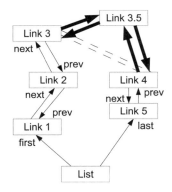

FIGURE 18.3: The changes needed to insert the number 3.5 into the list.

Here is a code example showing how to insert some items at the beginning and end of a list using `push_front` and `push_back`. We then print out the list.

```
// use a list to store items in priority order
list<string> list;
list.push_back("Drinking");
list.push_back("Dancing");
list.push_front("Exam");
list.push_front("Revision");

std::cout << "Todo list\n";
for (auto item : list) {
    std::cout << "Item " << item << "\n";
}
```

314   *C++ for Financial Mathematics*

This then prints out:

```
Todo list
Item Revision
Item Exam
Item Drinking
Item Dancing
```

Suppose we now wish to insert an item before "Exam". The class `list` has a function `insert` but it takes two parameters. The first is an iterator which must be pointing to the point where we want to insert, the second is the data we want to insert. Therefore we will first want to iterate through the list up to the desired point and then call insert. The code is below.

```cpp
auto i = list.begin();
while (i != list.end()) {
    if (*i == "Exam") {
        list.insert(i, "Dentist");
        break;
    }
    i++;
}

std::cout << "Todo list\n";
for (auto item : list) {
    std::cout << "Item " << item << "\n";
}
```

As expected, this code will print out the same list but with the item "Dentist" inserted before "Exam".

Naturally you will think that this code is longer than is necessary and will want to write a helper function to make it easier to insert elements. In fact, C++ already contains a function `find` which makes this code fairly easy.

```cpp
auto iter = find(list.begin(), list.end(),"Exam");
list.insert(iter, "Dentist");
```

The function `find` will search through a range of values looking for a specific item and returns an iterator to that point. To specify the range of values you provide an iterator that points to the beginning of your search region and an iterator that points to the end (i.e., one after the last element).

The function `find` is defined in the library `<algorithm>`. The class `list` is defined in `<list>`.

`<algorithm>` contains a number of helpful functions to perform tasks such as remove specified items from a container, sort the elements of a container, find the smallest element in a container, and so forth.

The Standard Template Library 315

18.8 The container `initializer_list`

A new container type called an `initializer_list` was introduced in C++11. These are useful because they are easy to create and populate with a list of values. They are not intended for long-term storage of data. Here is an example of an initializer list being used to create a list of doubles and then print them out.

```
std::initializer_list<double>
    list = { 1, 2, 3, 4 };
for (auto d : list) {
    std::cout << "Value "<<d<<"\n";
}
```

Initializer lists are read-only data structures, so they are primarily used as a constructor parameter to another container. For example, if you want to create a vector that contains some fixed specified values, you could use code such as that shown below.

```
std::vector<double> v({ 1, 2, 3, 4 });
```

This code creates a vector which contains the values 1, 2, 3, and 4. This can now be used like any other vector.

Most collection classes have a constructor that takes an `initializer_list` as a parameter so that this convenient notation can be used.

18.9 The containers `map` and `unordered_map`

So far our containers have simply stored a sequence or a set of values. Another broad class of data structures is used to store mappings from a key to some associated data. To give an example, a dictionary can be thought of as a data structure which maps a key (a word) to a value (the meaning of the word).

Two classes that can be used to store such mappings in C++ are the class `map` and `unordered_map`. They can be found in `<map>` and `<unordered_map>`, respectively.

The two classes are almost indistinguishable in how you use them. The key differences are:

(i) The elements in a `map` are ordered by the keys, just as the entries in a dictionary are ordered;

316   *C++ for Financial Mathematics*

(ii) `unordered_map` is usually faster than `map`.

Thus you should usually use `unordered_map` unless the ordering is important to you. Note that the items you use as keys in a `map` must overload < so that the map knows how to sort them in the correct order.

Here is how you can insert data into a map and then iterate through it to print the contents.

```
map<string, string> fruitToCol;
fruitToCol["apples"] = "green";
fruitToCol["bananas"] = "yellow";
fruitToCol["plums"] = "purple";
fruitToCol["oranges"] = "orange";
for (const pair<string,string>& p : fruitToCol) {
    cout << "The color of ";
    cout << p.first;
    cout << " is ";
    cout << p.second;
    cout << "\n";
}
```

You will see that as we iterate through the map, we obtain objects of type `const pair<string,string>&`. Each entry in the map is a key-value pair. The first entry is the key, the second entry is the value. We would usually write `auto` for the type of `p` rather than write everything out in so much detail.

The key advantage of the `map` structures are that we can very quickly obtain the value associated with a given key.

```
auto i = fruitToCol.find("plums");
cout << "Plums are " << (i->second)<<"\n";
```

In this example, we happen to know that the key "plums" is indeed stored in the map so we can be sure that i will be a valid iterator and hence the call i->second will contain the colour of a plum (purple). Sometimes you do not know whether a key is stored in the map or not. The code below tests if a key is in the map and acts accordingly.

```
string fruit = "jackfruit";
auto iter = fruitToCol.find(fruit);
if (iter == fruitToCol.end()) {
    cout << "The color of " << fruit;
    cout << " is unknown\n";
} else {
    cout << fruit <<" are "<< (i->second) << "\n";
}
```

So, to see if an element exists in the map, we compare the iterator returned by find with the iterator returned by end.

Although all our examples have used map, the code for an unordered_map would be identical.

18.9.1 How a map works

A map typically stores its data in a tree structure consisting of a number of nodes. Each node stores a key, a value, and a pointer to the nodes to the left and to the right. Where there is no adjacent node, a null pointer is used to indicate this. This is illustrated in Figure 18.4 where the black circles indicate the null pointers.

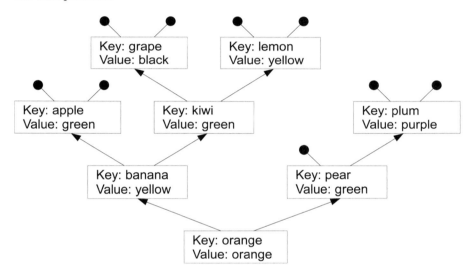

FIGURE 18.4: A possible arrangement for the nodes in a map.

The map class ensures that at all times the key of a node is always greater than the key of the nodes on its left and less than the key of the nodes on its right.

This means that when we wish to search for the data associated with a given key, it is easy to quickly navigate up the tree, taking either a left branch or right branch according to whether the key of the node you are currently looking at is greater than or less than your key. Typically you would expect the depth of the tree to have a height of $O(\log_2(N))$ and so it should only take $O(\log_2(N))$ steps to lookup an element with a given key.

Inserting data and removing data is a little more tricky than searching for data. When you insert and remove nodes you need to make sure that the ordering property is preserved. You also want to make sure that you keep the depth of the tree of order $\log_2(N)$ at all times. There are standard algorithms

318 *C++ for Financial Mathematics*

to do this. The red–black tree algorithm is one well-known approach. The details are not important for this book but can be found in any book on algorithms and data structures.

The end result is that a `map` can be used to search for data in $O(\log(N))$ and you can insert and remove data in $O(\log(N))$.

As should be clear from this discussion, a `map` depends crucially on the ordering of the keys. You must have in implementation of < when using a `map`, or else you must tell the `map` what function to use instead.

The implementation of a `set` is essentially the same as a `map` except that there is no need to store a `value`. This explains the performance characteristics of the class `set` that we mentioned in Section 18.5.

18.9.2 How an `unordered_map` works

Let us suppose that you wish to store approximately 100 key-value pairs so that we can look up the value given the key as quickly as possible. If you could somehow quickly associate a unique number between 0 and 99 to each key, then you could store your data in a `vector` instead of a `map`. Simply take a key, find the associated number, then read the element at the given index in the vector.

The problem is that we can't associate a unique number to a key in a straightforward way. However, one idea that *almost* works for a string is to take all the character codes in the string and add them up. This will give you a number for each string which we will call the *hash code*. The hash code is almost unique in the sense that two random strings aren't very likely to have the same hash code.

If you then take the remainder after dividing the hash code by 100, you will have a number between 0 and 99. Let us call this the *bucket id* of the string.

We can now group all the keys we want to use into *buckets*. Two strings are in the same bucket if and only if they have the same bucket id.

In an unordered map, one stores a vector of bucket objects. There is a bucket for each bucket id. The bucket objects contain a list of all the key-value pairs where the key happens to have the given bucket id. The situation is depicted in Figure 18.5. This shows an unordered map with 10 buckets, so the bucket id has been computed by taking the remainder on dividing the hash code by 10. In other words, the bucket id is just the last digit of the hash code. The hash codes have been computed by adding up the character codes for each letter in the key as described above.

To search for an element in an unordered map given a key, you proceed as follows.

1. Compute the hash code of the key.

2. Compute the bucket id by dividing the hash code by the number of buckets and taking the remainder.

The Standard Template Library

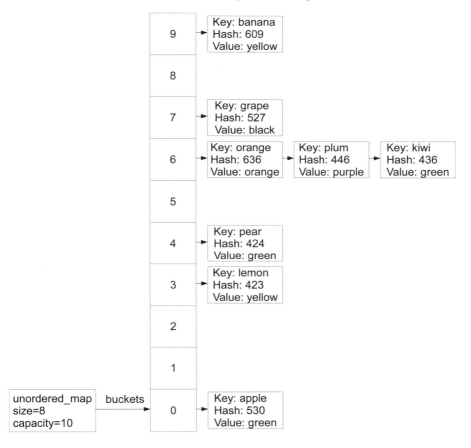

FIGURE 18.5: An unordered map with 10 buckets.

3. Read off the correct bucket from the vector of buckets.

4. Run through all the key-value pairs in the bucket until you have found the correct element.

The idea is that there will typically only be a very few elements in each bucket since the hash code for each key is essentially random. If we need to store N items, we can use an unordered map with $O(N)$ buckets so that there will still be only a handful of items in each bucket. This means that we can look up an element in a time which is roughly $O(1)$ as against the time $O(\log(N))$ required for a map.

Inserting new elements and removing elements is similar. One finds the correct bucket and inserts or removes the elements from the list. The only additional complication is that if the size of the unordered map increases over

320   *C++ for Financial Mathematics*

a certain amount, one should increase the number of buckets and recompute all the bucket ids. This process is called re-hashing.

Thus, roughly speaking, an unordered map allows one to insert, remove, and query data in approximately $O(1)$ and so is often faster than a `map`.

The key functions an unordered map needs to be able to perform on a key are to compute the hash code and to check if two keys are equal. For strings and numeric types, the unordered map will use a sensible choice of hash code and will test equality using `==`. For user-defined types, the default behaviour is that two keys are considered equal if and only if they are precisely the same instance of the object. If you wish to use a more refined notion of equality for your keys, then you should consult the documentation for `unordered_map`.

Note that the actual hash code used for a string in C++ is computed using a slightly cleverer algorithm than just adding up all the character codes in a string that is more likely to produce unique hash codes.

18.10   Storing complex types in containers

Tip: Store large objects using shared pointers

It is better to store large objects in containers using a `shared_ptr`. For example it is better to use a `vector< shared_ptr<Matrix>>` to store a list of matrices than a `vector<Matrix>`.

It is important to know that when you store data in a C++ container, the container stores a copy of the data. Similarly, when you extract data from a container, you obtain a copy of the data.

This is reasonable behaviour for small data types such as `int` or `double` or small `string` objects. However, if you are storing large objects in containers this will be very inefficient. The solution is to always store large objects using a `shared_ptr`.

18.11   A mathematical model for multiple stocks

STL classes will be invaluable in generalising our `Portfolio` pricing code to a model containing multiple stocks. Before describing the computational aspects of this problem, let us briefly describe a mathematical model that can be used for such a market.

The Standard Template Library 321

Our model will be simply the most obvious multi-dimensional generalisation of Brownian motion. It is a natural first attempt at writing a multi-dimensional model but is by no means the final word on modelling more complex stock markets.

We assume that the changes in the stock prices over each time interval δt are determined by R independent normally distributed risk factors ϵ^j with j varying between 1 and R. We write $\underline{\epsilon}_t$ for the R-vector of risk factors at time t.

We have n stock prices S_t^i at each time t with i between 1 and n. We write $z_t^i := \log(S_t^i)$ and write $\underline{z}_t$ for the vector of log stock prices at time t.

Our model is that the log of the stock prices evolves according to the difference equation:

$$\underline{z}_{t+\delta t} = \underline{z}_t + \underline{\eta}\delta t + (\delta t)^{\frac{1}{2}} L\underline{\epsilon}_t \qquad (18.1)$$

where $\underline{\eta}$ is an n vector and L is an $n \times R$ matrix. It is not significantly more difficult to simulate this multi-dimensional model than it is to simulate a one-dimensional model.

Given the value at z_t, the value at of z at time $z_{t+\delta t}$ will have a multi-dimensional normal distribution with covariance matrix $(\delta t) L L^\top$ where the superscript $\top$ denotes the matrix transpose. We write $A = L L^\top$ for the covariance matrix over a year. Two models of the form (18.1) with the same $\underline{\eta}$ should be considered equivalent if they have the same value for the covariance matrix A.

Normally one chooses the matrix L for this model by first choosing a covariance matrix A and then finding a matrix A such that $L L^T = A$. Such a matrix is called a *pseudo square root* of A. An algorithm for finding a pseudo square root of a covariance matrix called "Cholesky decomposition" was given in Exercise 16.7.5.

The larger the value of R one uses, the longer it takes to compute each term in the model (18.1). Computing each successive term in (18.1) requires $n \times R^2$ calculations. For this reason when one has a large number of stocks one might try to model the market with less risk factors than stocks. To do this, one works with a lower dimensional approximate pseudo square root of the covariance matrix instead of a true pseudo square root. This can be done by finding a basis of eigenvectors for the covariance matrix and using only the eigenvectors with the largest eigenvalues to create the matrix L. This allows one to use a moderate number of risk factors to model a large market.

For simplicity, however, we will assume from now on that L is the Cholesky decomposition of a given covariance matrix A. This is the situation we shall model in our software.

If we consider the behaviour the i-th component of our log stock price vector $\underline{z}_t$ we see that it is normally distributed with variance $(\delta t) A_{(i,i)}$ and mean $(\delta t)\underline{\eta}_i$. Thus each stock individually follows a geometric Brownian motion with drift $\underline{\eta}_i + \frac{1}{2}A_{(i,i)}$ and volatility $\sqrt{A_{(i,i)}}$.

We deduce that this is a risk-neutral model for stock prices in a world with

322   C++ for Financial Mathematics

fixed interest rate r if and only if

$$\underline{\eta}_{(i)} = r - \frac{1}{2} A_{(i,i)}.$$

In addition we see that the vector $\underline{\eta}$ can be computed from the individual drifts of each stock.

18.12 Using the Standard Template Library in FMLib

We wish to enhance FMLib so that it is possible to price a portfolio which contains options on more than one stock. In addition we would like to be able to price options such as a Margrabe option where the contract involves the interaction of two stocks (see Exercise 18.12.2).

The key changes we need to make to FMLib are summarised in Figure 18.6. Let us also explain them in words.

The `ContinuousTimeOption` interface is changed so that its `payoff` function now takes a `MarketSimulation` object as a parameter. The previous version of this function took a matrix of stock price paths as a parameter.

The `MarketSimulation` class is a simple class that stores a map between the name of a stock and a matrix of stock price paths. It stores all our simulations of the stock market. In order to compute the payoff of an option on multiple stocks, one queries the `MarketSimulation` for the price paths for each of the stocks involved in the option contract. One can now compute the payoff of the option according to the option contract.

In addition to changing the `payoff` function, the `ContinuousTimeOption` now has a method `getStocks` which can be used to query which stocks the option depends upon. To ensure that this method does not return duplicates, its return type is `set<string>`.

Most options involve only a single stock, so our base class `Continuous-TimeOptionBase` has been enhanced so that it stores the name of a single stock. It implements `getStocks` by returning this name. `Continuous-TimeOptionBase` also implements the `payoff` function that takes a `Market-Simulation` as a parameter by extracting the matrix of stock price paths for this single stock and then calling an abstract function, which is also called `payoff` but which takes a single `Matrix` as parameter. Since all our existing subclasses of `ContinuousTimeOptionBase` already implement this second version of the `payoff` function, they will now automatically implement the modified `ContinuousTimeOption` interface.

To generate the `MarketSimulation` objects we use a class called `Multi-StockModel`. This class represents the mathematical model for multiple stocks described above. It has functions `generatePricePaths` and `generateRisk-`

The Standard Template Library

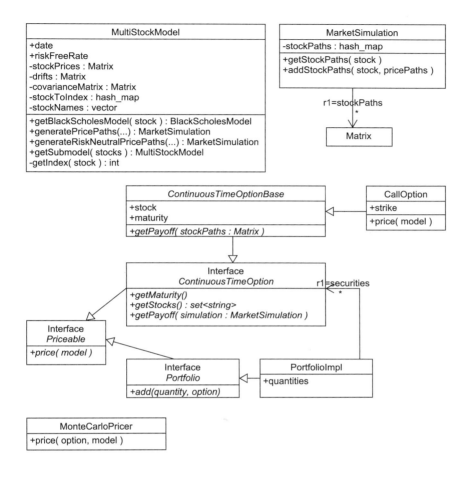

FIGURE 18.6: UML summary of changes to FMLib

`NeutralPricePaths` that create `MarketSimulation` objects according to the configuration of the model.

Internally the `MultiStockModel` stores most of its data using matrices. For example, the current stock prices are stored as $n \times 1$ matrix—i.e., a column vector. Each of these stock prices corresponds to the price of some stock in the market. The mapping between the names of the stocks and the index of the stock in this column vector is stored in two different ways. Firstly they are stored in an `unordered_map` called `stockToIndex` which maps the name of a stock to the index. Secondly the names are stored in index order in a vector called `stockNames`. Together these data structures allow one to quickly map between stock names and indices and vice versa. To help with this mapping,

MarketSimulation has a private function **getIndex** which maps a stock name to its index.

The other data in **MarketSimulation** are stored using matrices with the same indices. In particular there is a column vector of drifts and a covariance matrix that describes both the volatilities of all the stocks and their correlations.

Because we will often want to price options and portfolios that do not involve the whole market, the **MultiStockModel** class has a **getSubmodel** function. This takes as parameter a list of stocks and returns a new **Multi-StockModel** which has been reduced in size to only involve the desired stocks. For convenience it also has a function to return a **BlackScholesModel** object given a single stock.

The **MonteCarloPricer** class can now price **ContinuousTimeOption** instances given a multi-stock model. It first creates an appropriate sub model using the **getStocks** function of the option and then the **getSubmodel** function of the model. It then uses this sub model to generate risk-neutral price paths. The option is able to compute its payoff for all of these price paths. The Monte Carlo price is then the discounted mean payoff.

Two other small changes worth mentioning are:

(i) The **Priceable** interface has been changed so the **price** method takes a **MultiStockModel** as a parameter. It used to take a **BlackScholes-Model**. All the implementations of this **price** method have been changed accordingly.

(ii) The **Portfolio** class now only allows you to add **ContinuousTimeOption** objects when previously any **Priceable** could be added. The reason is that this makes it easier to complete Exercise 18.12.3.

The actual implementation of most of the methods is not challenging. Let us show a few examples.

First, here is the implementation of **getIndex** in the class **MultiStock-Model**. We chose to inline this function. It simply uses the recipe given above for searching in a map. We have decided that we will simply report an error if the user passes an invalid stock code.

```
int getIndex(const std::string& stockCode)
        const {
    auto pos = stockToIndex.find(stockCode);
    ASSERT(pos != stockToIndex.end());
    int idx = pos->second;
    return idx;
}
```

The **getStockPaths** function on MarketSimulation is essentially identical.

```
SPCMatrix getStockPaths( const std::string& stock)
```

The Standard Template Library 325

```
        const {
          auto pos = stockPaths.find(stock);
          ASSERT(pos != stockPaths.end());
          return pos->second;
  }
```

One point that this code makes clear is that our `MarketSimulation` class stores matrices using `shared_ptr`. As discussed above, since containers copy the data that they store, it would be very inefficient to store a `Matrix` in a container directly. This code also demonstrates our convention of using a `typedef` to create a `shared_ptr` to a `const` with name beginning "SPC". You can see this convention being used in the return type of the function `getStockPaths`.

As a more complex example, here is the code required to create a sub model of a `MultiStockModel` given the names of the stocks you want included in the sub model. The code has to find the correct indices for these stocks and then extract the necessary submatrices from the member variables `drifts`, `stockPrices`, and `covarianceMatrix`.

```
MultiStockModel MultiStockModel::getSubmodel(
    set<string> stocks) const {

    int n = stocks.size();
    Matrix drifts(n, 1);
    Matrix stockPrices(n, 1);
    vector<string> newStocks(stocks.begin(),
                             stocks.end());
    Matrix cov(n, n);

    int newIndex = 0;
    for (auto& stock : stocks) {
        int idx = getIndex(stock);
        drifts(newIndex) = this->drifts(idx);
        stockPrices(newIndex) =this->stockPrices(idx);
        newIndex++;
    }

    int i = 0;
    for (auto& stockI : stocks) {
        int j = 0;
        for (auto& stockJ : stocks) {
            int oldI = getIndex(stockI);
            int oldJ = getIndex(stockJ);
            cov(i, j) = covarianceMatrix(oldI, oldJ);
            j++;
        }
```

326   *C++ for Financial Mathematics*

```
            i++;
        }
        MultiStockModel ret(newStocks, stockPrices,
                            drifts, cov);
        ret.setDate(getDate());
        ret.setRiskFreeRate(getRiskFreeRate());
        return ret;
}
```

This code demonstrates the convenience of looping using the special **for** syntax for containers. One point worth noticing is that we can construct a new vector that contains the data in an existing collection as follows:

```
        vector<string> newStocks(stocks.begin(),
                                 stocks.end());
```

Container classes usually have constructors that take two iterators and copy all the data between the iterators into the newly created container. Nine times out of ten, one uses this to copy all of some existing container, but it can be useful occasionally to only copy a part.

Finally we note that the class `MultiStockModel` generates price paths using the model described in Section 18.11. The code is not much more complicated than the code required to simulate a single stock price.

Exercises

18.12.1. In Exercise 3.9.3 we saw that recursion was an inefficient approach to computing the Fibonacci numbers. However, if we use a *cache* of values, recursion can be efficient. A cache is simply a data structure where you store the results of previous computations. You could use a global variable containing a **map** to cache values of the Fibonacci sequence that you have already computed. Implement this idea and confirm that it outperforms recursion without a cache.

18.12.2. Write an implementation of `ContinuousTimeOption` that represents a *Margrabe option*. This is an option on two stocks S^1 and S^2. At maturity T it pays off an amount equal to $\max\{S_T^1 - S_T^2, 0\}$. Confirm that you can price it by Monte Carlo and that is price matches the analytical formula given in [10].

18.12.3. Add a new method `priceByMonteCarlo` to the class `Portfolio` which groups together options by their maturity and then prices options with the same maturity using a single set of simulations. To do this you will want

The Standard Template Library 327

to write an implementation of `ContinuousTimeOption` that represents a portfolio of options on a variety of stocks that all have a single maturity.

In practice you might want to use a sophisticated algorithm that groups together only those options that require Monte Carlo pricing and prices those that can be handled analytically as before. You might also want to produce a single simulation of stock prices across all the maturities in your portfolio.

18.12.4. Write a templated function `mapGet` which gets a value from a map given a key. If the value is not present in the map it should throw an exception. Your function should work equally well with `map` and `unordered_map`.

18.12.5. Write your own simple version of a hash map. You should only implement a function `store`, which allows you to store a key–value pair, and `get`, which returns the value associated with a key or else throws an exception. You can assume that both the key and values are strings. You should also only use a fixed number of buckets. You may use the `vector` and `list` classes, but you may not use `map` or `unordered_map`. Check that your hash map performs as expected.

18.13 Summary

The standard template library is a powerful tool which will allow you to write more sophisticated programs with ease. Key points we have covered are:

- The `typedef` keyword allows us to abbreviate complex type names. A convention for naming `shared_ptr` types is particularly convenient.

- The `auto` keyword allows you to avoid typing the full name of a class when the compiler can deduce it for you. It is possible to combine `auto` with `const` and `&`.

- Classes can contain member types. This is particularly useful when writing template algorithms.

- A container is any class that stores data and returns iterators when you call `begin` and `end`. There is a special syntax for looping through containers using `for`.

- C++ contains numerous container classes that make it easy to store data. They can also act as building blocks for implementing efficient algorithms. All containers behave in a broadly similar manner. There are clear conventions for naming functions such as `size`, `insert`, `remove`, `begin`, and `end`.

328   *C++ for Financial Mathematics*

- The library `<algorithm>` contains a number of functions that are very useful for working with containers such as `find`.

- Different data structures have different performance characteristics. You should choose your data structure based on how you intend to access and update the data.

- You should not store large objects in containers. Store them by reference using `shared_ptr` instead.

We should add that although the standard template library is very powerful, it is also rather difficult to learn. Unfortunately it makes heavy use of sophisticated ideas such as templates and operator over-loading. The syntax chosen for iterators also assumes that you are very used to working with pointers and so will find the use of `*` and `++` to be natural and convenient.

When the standard template library was first invented it was rather experimental. Time has shown that the object-oriented programming style is quite intuitive. For this reason many computer languages include some form of object orientation. On the other hand, the standard template library is quite difficult to grasp for new programmers. As a result, the use of templates has not been adopted in many other languages.

Tip: Use the standard template library, but don't copy its design.

You should know how to use the standard template library as it contains invaluable tools. Nevertheless, you should not rush to copy the design of the standard template library. See if you can find an object-oriented solution to your problem before using templates.

Chapter 19

Function Objects and Lambda Functions

In Section 10.6 we designed an interface `RealFunction` to represent the mathematical notion of a function mapping real numbers to real numbers. This solved the problem of writing an integration routine that can integrate any real function.

However, there is a better solution in C++, called writing *function objects*. A function object is simply an instance of a class that overloads the operator (). Function objects are also sometimes called *functors*. The big advantage of using function objects is that there is a special syntax for writing function objects called *lambda functions*. This makes it quicker to write a function object than a normal class.

19.1   Function objects

Here is an example of an integration function which takes as a parameter a function object `f`. This code simply repeats the function we used in Section 10.6 to integrate using the rectangle rule, but this time we are using a function object rather than the interface `RealFunction`.

```
double  integrate (
        function<double(double)> f,
        double a,
        double b,
        int nSteps) {
    double total = 0.0;
    double h = (b - a) / nSteps;
    for (int i = 0; i<nSteps; i++) {
        double x = a + i*h + 0.5*h;
        total += h*f(x);
    }
    return total;
}
```

329

330   *C++ for Financial Mathematics*

The first thing to notice is that the type of **f** is declared as

```
function<double(double)> f
```

Here **function** is a class defined in the library **<functional>** and in the namespace **std**. The interesting bit is the template type. This example means that **f** represents a function that returns a double and takes a single **double** as a parameter. If you wish to use function objects with different return and parameter types, the general syntax is:

```
function<ReturnType(ParameterType1,ParameterType2,...)> f
```

Declaring the type of the function is the only complex part of using function objects. When we use the function object **f** it is just the same as using an ordinary function because it has overridden the operator (). Here is the line of code where **f** is used:

```
total += h*f(x);
```

To write a function object, you must write a class which overloads the () operator. Here is an example of a class for evaluating the mathematical function $\sin(x)$.

```cpp
class SinFunction {
public:
    double operator()(double x) {
        return sin(x);
    }
};
```

Putting everything together, here is a test that our integration function works as expected.

```cpp
void testIntegrateSin() {
    SinFunction integrand;
    double value = integrate(integrand, 0, 1, 1000);
    ASSERT_APPROX_EQUAL(-cos(1.0) + cos(0.0),
        value, 0.01);
}
```

19.2   Lambda functions

The big advantage of using function objects rather than writing your own interface is that you can use *lambda functions* to write function objects very quickly.

Function Objects and Lambda Functions 331

As an example, suppose that we want to write a function that takes parameters a, b and c and then computes

$$\int_0^1 (ax^2 + bx + c)\, \mathrm{d}x$$

using the trapezium rule. We will first need to write a function object to compute the integrand $ax^2 + bx + c$. We can then compute the integral using our `integrate` function.

Here is the long answer which doesn't use lambda functions:

```
class QuadraticFunction {
public:
    /* Members */
    double a;
    double b;
    double c;
    /* Constructor */
    QuadraticFunction(double a,
        double b,
        double c) :
        a(a), b(b), c(c) {}
    /* Operator */
    double operator()(double x) {
        return a*x*x+b*x+c;
    }
};

double integrateQuadratic(double a,
    double b,
    double c) {
    QuadraticFunction integrand(a, b, c);
    return integrate(integrand, 0, 1, 1000);
}
```

We have to define a class, give it appropriate member variables and a constructor, and override the operator (). We then need to construct an instance of this class and pass it to `integrate`. A lambda function allows us to do all of this in only a few lines.

```
double integrateQuadratic2(double a,
    double b,
    double c) {
    auto lambda =
        [a, b, c](double x) {
            return a*x*x + b*x + c;
        };
```

332   *C++ for Financial Mathematics*

```
    return integrate(lambda, 0, 1, 1000);
}
```

The variable `lambda` is still a function object, just as `integrand` was in the more verbose version of the code. The new syntax is the declaration of the lambda function:

```
        [a, b, c](double x) {
            return a*x*x + b*x + c;
        };
```

This is an extremely concise syntax which means:

- Generate a class, we don't care what it is called.

- We will want to *capture* the local variables a, b, and c of integrate-Quadratic2 and have them as member variables of our class. When writing a lambda function, you list the captured variables in square brackets.

- We want to write an overload of operator () that takes a single **double** parameter which we will call x. When writing a lambda function, you list the parameter types and parameter names in round brackets.

- The actual computation for the function is written inside curly brackets and can use both the captured variables and the parameters.

In summary, the syntax of a lambda function is:

```
[CaptureParameters] (FunctionParameters) {FunctionImplementation}
```

The great thing about lambda functions is that as well as being quicker to write, they are easier to read (at least when you are used to the syntax). This is because all the verbose and distracting boiler plate code needed to define a new class is hidden.

There is a lot of flexibility in how you write the capture parameters.

(i) You can specify that you would like to capture local variables by reference, by using the & symbol before the parameter name.

(ii) You can specify that you would like to capture all variables by reference simply by specifying just &.

(iii) You can specify that you would like to capture all variables by value, by specifying just =.

(iv) If your lambda function is written inside a member function of a class, you can capture the member functions and member variables of that class by specifying **this**.

Function Objects and Lambda Functions 333

As an example, here is how to integrate the payoff of a path-independent option between the limits a and b.

```
double integratePayoff(PathIndependentOption& o,
                       double a,
                       double b) {
    auto lambda =
        [&o](double x) {
        return o.payoff(x);
    };
    return integrate(lambda, a, b, 1000);
}
```

Since the class `PathIndependentOption` is abstract, we must capture the variable o by reference. This is why the list of capture parameters says we are capturing &o.

This same problem was considered in Section 10.6. As you can see, the use of lambda functions makes the solution much shorter.

Notice that when you store capture parameters by reference, you must always be careful to ensure that function object will be deleted before any of the parameters it has captured. This is for precisely the same reason as one needs to ensure that an instance of a class is deleted before any member variables that are stored by reference.

19.3 Function pointers

It is possible to pass an actual function as a parameter, it is not always absolutely necessary to write a function object. For example, to integrate

$$\int_0^1 (x^2 + 2x + 1)\mathrm{d}x$$

we can write an ordinary function representing the integrand as follows:

```
static double integrand(double x) {
    return x*x + 2 * x + x;
}
```

We can then pass this integrand to our integrate function as follows.

```
double testIntegrateFunctionPointer() {
    double value = integrate(&integrand, 2, 1);
    ASSERT_APPROX_EQUAL(
        value,
```

334   *C++ for Financial Mathematics*

```
        2.3333, 0.01);
}
```

Note the & symbol before `integrand`. `&integrand` is called a *function pointer*.

The C language does not have object-oriented features. In particular you can't write function objects. The closest you can get is to write functions and pass them round with function pointers. If you use libraries written in C you will see that they often use function pointers. However, C++ programmers usually prefer to write their libraries using function objects because they are both more powerful and more convenient.

19.4   Sorting with lambda functions

We have used lambda functions to represent a real valued function defined on the reals. However, lambda functions can be used for many different kinds of function, not just mathematical ones. We will give one example in this section that shows how lambda functions can be used to customise the sort order of the `sort` function.

By default, the < function for strings is based on ASCII character codes so "A" < "B" < ... < "Z" < "a". This means that using the `sort` function directly on a vector of strings will not put the list in alphabetical order. All the strings that start with upper case will come before all the strings that start with lower case. Fortunately, `sort` allows you to specify a comparison function that it will use to decide whether x should come before y or not.

```
void sortExample() {
    vector<string> list({ "Z", "x", "a", "B" });
    sort(list.begin(), list.end(),
        [](string& x, string& y) {
            return uppercase(x) < uppercase(y);
        });
}
```

In this example, we create a lambda function that compares two strings x and y and returns whether x comes before y alphabetically. To do this, it simply converts each string to upper case and then compares them. The function `uppercase` is not built into C++, it is a helper function we have written for this purpose.

C++ libraries often make heavy use of function objects. Using them is essential if you want to get the most out of C++.

Exercises

19.4.1. Using a lambda function, compute

$$\int_0^\pi \sqrt{1 + \sin^2(x)}\,dx.$$

19.4.2. Write a function to compute the price of an option that has an arbitrary payoff $f(S)$ at maturity T in the Black–Scholes model. Your code should be implemented by computing the expected payoff by integration using Equation (A.3).

19.4.3. Suppose that $f : [a, b] \to \mathbb{R}$ is a continuous function and we wish to solve the equation $f(x) = y$ for some value of y lying between $f(a)$ and $f(b)$. Obviously there is a solution and it must lie in one of the two sub-intervals $[a, (a + b)/2]$ or $[(a + b)/2, b]$. Depending on the value of f at $(a + b)/2$, we can determine which of the two sub-intervals we need to look in to solve the equation: Simply choose the interval such that y lies between the values of f at the end points of the interval.

This gives a recursive algorithm to numerically solve the equation $f(x) = y$ within a given tolerance ϵ. To solve the equation on $[a, b]$:

(i) Check if either of the end points is a solution within the desired tolerance, in which case we are done.

(ii) Otherwise, decide which sub interval $[a, (a + b)/2]$ or $[(a + b)/2, b]$ must contain the solution. Use recursion to apply the algorithm to the smaller interval.

This algorithm is called the method of bisection. Implement it. (There are many other algorithms to solve equations of this type, such as the Newton–Raphson method. You can find implementations in libraries such as GSL, the GNU Scientific Library.)

19.4.4. It is easy to calibrate all the parameters in the Black–Scholes formula to the market, except for the volatility σ.

For example, the original stock price is usually taken to be the average of the bid and ask prices quoted on the market for buying and selling the stock. Similarly, the risk-free rate can be taken to be some specific market interest rate, for example the LIBOR rate.

However, the volatility parameter does not correspond to any market data in such a clear manner. Instead, one often attempts to calibrate the Black–Scholes model to the market by inverting the Black–Scholes formula and using the prices of call options on the market to deduce the volatility. The volatility computed in this way is called an *implied volatility*.

336   *C++ for Financial Mathematics*

Use your answer to the previous exercise to write a function `implied-Volatility` that:

- Takes parameters S, r, K, T and the price of a call option with strike K and maturity T
- Computes the corresponding value of σ such that Equation (A.6) holds.

19.4.5. Find the current market data for call options on the S&P 500 Index and plot a graph of how the implied volatility changes with the strike price. If the Black–Scholes model was perfectly accurate, the implied volatility would be independent of the strike price. Since the Black–Scholes model makes a number of unrealistic assumptions, you will instead find that the implied volatility varies with the strike price. This demonstrates that while the Black–Scholes model is useful, it is not accurate enough to replicate actual market prices and more sophisticated models must be used.

19.5 Summary

Passing functions as parameters is a common requirement in C++. Use the class `std::function` to pass functions as parameters. This allows you to use lambda functions to quickly create new functions as needed.

Chapter 20

Threads

A computer is capable of doing more than one thing at once. As well as running more than one program at a time, it is also possible for each program to be performing more than one task at each time. Each separate program that is running is called a *process*. Within each process, each separate running task is called a *thread*. The programs we have written so far contain a single thread. This chapter is about writing programs that execute multiple threads at once.

If a computer has multiple CPUs, each CPU can manage its own thread of execution and so one can run multiple threads simultaneously. Modern processors often contain multiple *cores*. You can think of each core as being effectively a processor in its own right. Each core can execute a separate thread.

However, even if you only have one processor you can still run multiple threads. The threads will appear to run simultaneously, but in fact the operating system will simply allocate a little bit of time to each thread in turn. More generally, you can run more threads than processors. One of the main tasks of the operating system is to determine which threads to run at any one time. For example, the operating system may give priority to graphical user interfaces and less priority to background computations.

A multi-threaded program is simply a program that contains more than one thread. Here is a short list of the possible benefits one might hope to obtain from writing a multi-threaded program.

(i) **CPU Performance:** In order to use all your processors or cores you must run multiple threads, so multi-threading is essential to making full use of your computer resources.

(ii) **Network Performance:** When you are waiting for network operations to complete (e.g., a web page to download), you might as well use your processors to do other things.

(iii) **Fairness:** If you are writing a web server, users who want to view one page shouldn't have to wait half an hour for some other user's enormous download to complete.

(iv) **Usability:** While a program is performing a lengthy calculation, users appreciate being able to perform other tasks while the processing happens "in the background".

338   *C++ for Financial Mathematics*

(v) **More natural programming:** *Very occasionally* a program will be simpler to write if you use multiple threads.

However, there is a price to be paid for all these positive benefits. Here are some of the pitfalls one must consider before deciding to write a multi-threaded program.

(i) **Complexity:** Multi-threaded code is much harder to write and understand than single-threaded code.

(ii) **Testability:** Multi-threaded code is much harder to test than single-threaded code.

(iii) **CPU Performance:** Using multiple threads can actually slow your computer down if the code is not written well. This is because there is some overhead in sending data between threads. If your code spends most of its time moving data from one thread to another it will perform badly.

(iv) **Difficulty:** The human brain seems to struggle with parallel processing. We're all quite hopeless at multi-tasking and we're all quite bad at writing multi-threaded code.

20.1 Concurrent programming in C++

20.1.1 Creating threads

To create a thread one constructs a `thread` object, passing in a pointer to the function to call when the thread executes and any parameters required by the function. `thread` is defined in the library `<thread>`.

The code below contains a function `primalityTest` that takes a single parameter, which is a pointer to an `int`. The function `primalityTest` simply tries every possible divisor of the referenced `int` and sees if it can find a factor. If it finds a non-trivial factor, it prints out that the referenced `int` is not prime. Otherwise it prints out the fact that the referenced `int` is a prime.

```
void primalityTest( int* pointerToInt ) {
    int toTest = *pointerToInt;
    for (int i=2; i<toTest; i++) {
        if ((toTest % i)==0) {
            INFO( toTest << " is not prime" );
            return;
        }
    }
}
```

```
      INFO( toTest << " is prime" );
}

void testPrimes() {
    int values[3] = {1299817,1299821,1299827};
    thread t1( &primalityTest, &values[0] );
    thread t2( &primalityTest, &values[1] );
    thread t3( &primalityTest, &values[2] );
    t1.join();
    t2.join();
    t3.join();
}
```

We wish to run the function `primalityTest` using three separate threads. The function `testPrimes` achieves this by creating three `thread` objects. As you can see we use the expression `&primalityTest` to obtain a pointer to the primality test function.

When we have created each thread we wait for them to complete by calling `join`. The `join` function of a thread *blocks* (i.e., performs no processing) until the thread that has been joined completes.

20.1.2 Mutual exclusion

To see the problems that may arise when writing multi-threaded code, consider what happens if two threads simultaneously execute the code

```
cout << "The answer to calculation "<<i<<" is "<<j<<"\n";
```

The first thing to notice is that this code is actually equivalent to the code:

```
cout << "The answer to calculation ";
cout << i;
cout <<" is ";
cout <<j;
cout <<"\n";
```

If two threads execute this code at once, there is no guarantee which thread will be able to print out a given line of the message first. As a result the messages that are printed out may be superimposed confusingly. Below is an example of the type of output that may occur.

```
The answer to calculation The answer to calculation 1 2 is 19
is 18
```

These problems are very visible when one writes data to `cout`, but subtle problems can easily occur when multiple threads access the same data or the same objects. Consider for example the following code:

```
bool debitAccount(Account& account, double amount) {
    if (account.balance >= amount) {
        account.balance -= amount;
        return true;
    }
    return false;
}
```

To see why this code is dangerous, suppose that initially the account contains $100. Suppose that two threads both attempt to debit $100 from the account.

It is possible that, due to bad luck, it may happen that both threads will simultaneously check the balance and see there are sufficient funds. What may now happen is that one thread changes the balance to zero, and the second thread then changes it to −$100. The end result will be that the account will go overdrawn, even though it at first sight may look as though our if statement will prevent this occurring.

This is an example of a *race condition*. A race condition is any situation where two threads attempt to access the same resource and the outcome depends upon the order of events. In our example, one imagines that the two threads are "racing" to debit from the account.

Race conditions are inevitable unless we find some way of ensuring that two threads cannot read and write to the same data at the same time. *Mutual exclusion locks* provide a way of ensuring this.

The class `mutex` represents a mutual exclusion lock. It has two key methods, `lock` and `unlock`. We say that when the function `lock` returns, the thread has *acquired* the lock. When `unlock` is called, the thread has *released* the lock.

Only one thread can acquire the lock at a time. If thread A acquires the lock and then thread B tries to call `lock`, then call to `lock` won't return until thread A has called `unlock`. This guarantees that only one thread ever holds the lock. It is called a *mutual exclusion lock* because each thread can exclude another thread from running by holding the lock. They mutually exclude each other from running.

In the novel *Lord of the Flies*, a group of boys are stranded on a desert island. The boys find a conch shell on a beach and impose a "rule of the conch" on themselves. They agree that no one can speak unless he's holding the conch. This means that the boys are using the conch as a mutual exclusion lock to prevent them from talking over each other.

Here is an example of how we can use a mutual exclusion lock in a class called `Account` which represents a bank account. We give it a member variable which is a mutex and use this to ensure that no two threads can simultaneously debit from the same account:

```
bool debitAccount(Account& account,
                  double amount) {
    account.mtx.lock();
```

```
    bool ret = false;
    if (account.balance >= amount) {
        account.balance -= amount;
        ret = true;
    }
    account.mtx.unlock(); // don't do this
    return ret;
}
```

While this code works, you should not actually use the lock and unlock methods of a mutex directly. The reason for this is that if you lock a mutex you will need to ensure that unlock is called eventually no matter what happens. In particular, even if an error occurs, unlock should still be called.

A mutex is an example of a resource that must be released after it is acquired, even if an error is thrown. It is far from the only example. Memory is also a resource that should also be released eventually. The same is true for database connections and locks on the file system.

The solution for all problems of this type is to create a class that manages the resource and which makes sure that it is released by using a destructor. Destructors are called when a local variable is deleted even if an error has occurred.

In the case of a mutex, the class we need to manage the mutex already exists and is called a lock_guard<mutex>. (mutex is not the only lockable resource. You can create other types of lock_guard to manage other types of lockable resource. This is why lock_guard is a template.)

A lock_guard<mutex> calls lock on a mutex in its constructor and then calls unlock in its destructor. In particular, this means that if one uses a lock_guard<mutex> as shown below, the mutex will be unlocked even if an error occurs.

```
bool debitAccount(Account& account,
                  double amount) {
    lock_guard<mutex> lock(account.mtx);
    if (account.balance >= amount) {
        account.balance -= amount;
        return true;
    }
    return false;
}
```

Tip: Resource acquisition is initialisation (RAII)

Whenever there is some clean-up that *must* occur at the end of your method,

342   *C++ for Financial Mathematics*

use an appropriate class with a destructor that performs the clean-up. This is summarised by saying that:

- resource acquisition is initialisation;

- resource release is deletion.

When we introduced destructors in Section 16.2, the resource we wished to manage was memory. In this section we are viewing holding a lock as another form of resource acquisition. In both cases there is a limited supply of the resource that needs to be managed: A computer has finite memory; only one thread can hold a lock.

If you do not use mutual exclusion locks, you have very little guarantee about the order in which code will be executed (or appear to be executed). Two problems one has to consider when one doesn't use mutual exclusion locks are: The compiler may decide to optimise the code by reordering certain sections; processors have caches that contain data that may not be written to main memory for some time. C++ does contain some classes to enable you to write multi-threaded code without using mutual exclusion locks, but this is a specialist subject beyond the scope of this book.

Danger!

The compiler can reorder your code as part of the optimisation process. If you don't use locks you have very little guarantee on what order code executes. In other words, if you don't use locks some very strange bugs will occur.

20.1.3   Global variables and race conditions

Whenever you use a global variable, there will be a potential race condition if your code is used in a multi-threaded program.

For example, the output streams `cout` and `cerr` are global variables and we have already seen that using these is problematic when writing multi-threaded code. In FMLib20 the macros `INFO` and `DEBUG_PRINT` have been updated to use a `mutex` so that messages do not overlap.

This is not the only use of global variables in our code. The function `randuniform` uses a global variable to store the current state of the random number generator. We must therefore add a mutex to the implementation of `randuniform` to prevent race conditions:

```
/*  MersenneTwister random number generator */
static mt19937 mersenneTwister;
/*  Mutex to protect static var */
```

Threads

```
static mutex rngMutex;

/* Reset the random number generator.
We ignore the description string */
void rng(const string& description) {
    ASSERT(description == "default");
    lock_guard<mutex> lock(rngMutex);
    mersenneTwister.seed(mt19937::default_seed);
}

/* Generate random numbers */
Matrix randuniform(int rows, int cols) {
    lock_guard<mutex> lock(rngMutex);
    return randuniform(mersenneTwister, rows, cols);
}
```

We have used encapsulation to ensure that the random number generator and the mutex can only be accessed from `matlib.cpp`. This allows us to ensure that it is impossible to use the random number generator without first locking the mutex.

20.1.4 Problems with locking

Using the `mutex` class may seem like a simple solution. However, in practice writing correct code for locking between threads can be rather difficult.

The first source of problems is that it is hard to notice that there is a problem if one omits the correct locking code. Often, one will find that the program appears to work most of the time and then intermittently fails. In many ways this is worse than simply failing outright. Should you trust the answers of a program that occasionally makes mistakes?

To make matters worse, writing unit tests doesn't help very much. Errors in locking occur when threads interleave in unexpected ways. It is hard to write tests that force this to happen.

In addition to these practical problems related to race conditions, one also has to consider the problem of *deadlock* that can occur once one uses multiple locks within a program.

To understand the meaning of a deadlock, suppose that we have two threads, A and B, and two mutexes a and b. Now suppose the following events occur in the given order:

1) Thread A locks mutex a.

2) Thread B now locks mutex b.

3) Keeping hold of mutex a, thread A tries to lock mutex b.

4) Keeping hold of mutex b, thread B tries to lock mutex a.

344 *C++ for Financial Mathematics*

5) Neither thread A nor thread B can proceed because they are each waiting for the other to complete.

This situation is called deadlock. It occurs whenever two or more threads are unable to proceed because they are mutually holding resources required by the other threads.

A famous example of deadlock is the problem of dining philosophers. The problem is usually expressed as follows:

- A group of philosophers sit in a circle round a table to eat noodles.

- There is a chopstick on the left and a chopstick on the right of each philosopher.

- Unfortunately they share chopsticks with their neighbours.

- Each philosopher decides at random moments to:

  - pick up a free chopstick on their left or right;
  - then pick up the second chopstick when it becomes available;
  - eat some noodles;
  - put down both chopsticks.

This situation is deadlock prone. For example, if every philosopher picks up the left chopstick, they will then find there is no chopstick on their right to pick up. The end result is that the philosophers will all starve to death.

A more prosaic example is given by transferring money between accounts. Consider the following method.

```
bool transferMoney(Account& from,
    Account& to,
    double quantity) {
    lock_guard<mutex> lock1(from.mtx);
    if (from.balance<quantity) {
        return false;
    }
    lock_guard<mutex> lock2(to.mtx);
    from.balance -= quantity;
    to.balance += quantity;
    return true;
}
```

This code attempts to ensure that we hold a lock on both accounts at the same time whenever we transfer money from one account to the other.

But suppose that two threads attempt to simultaneously transfer money between two accounts, account A and account B. Suppose that thread 1 transfers from A to B whereas thread 2 transfers from B to A. It is possible that

Threads 345

thread 1 will acquire the mutex on A, thread 2 will acquire the mutex on B, and then neither thread will be able to proceed.

There are various techniques one can use to try and resolve the problem. For example, you could:

(i) Write single-threaded code instead.

(ii) Only use one lock for all accounts.

(iii) Give the locks an ordering and insist that (say) lock A is always acquired before lock B.

(iv) Include a time-out so if you don't acquire a lock in a reasonable time frame you should release all the locks that you hold.

(v) Incorporate some deadlock detection and resolution algorithm in your locking classes.

All of these are sensible possible solutions to the deadlock problem. Which is the right choice depends upon what your priorities are in writing the system. For example, the first option is probably best if ease of programming is important to you and performance is unimportant.

Tip: Guidelines for multi-threaded code

- Don't use global variables. If you must have global variables other than constants you will need to use locks to protect them.

- Minimise the data shared between threads. The less that is shared, the less locking required.

- Where possible, use `const` data between threads as this won't require locking.

- Divide your code into simple sequential algorithms and small separate sections where threads communicate.

- Use standard established design patterns and classes for multi-threaded code.

- Don't write multi-threaded code unless there is a clear benefit. Even then only a tiny part of your code should involve threading.

For example, one important design pattern for writing multi-threaded code is to use a *database* to communicate between threads. Standard database software is specifically designed for concurrent use. Commercial databases support sophisticated "transactions" which are a way of allowing multiple users to perform complex interactions, with the illusion that they are using ordinary sequential code. The net result is that if you use a database to communicate between threads, then the database will do all the hard work so you don't have to. There is a downside. Communicating with a database is much slower than working in memory and using a mutex. Nevertheless, databases are great if you have distributed processes.

We won't pursue this design pattern further, but in practice it is one of the most important software design patterns. In this chapter we'll see some other multi-threading patterns that can be used entirely within one C++ program.

20.2   The command design pattern

Although it is possible to use the class `thread` directly, the *command* design pattern provides a more object-oriented alternative.

You can use the command design pattern whenever you have some "command" that you wish to instruct some other piece of code to perform on your behalf. For example, when writing a user interface you may wish to configure a button so that it performs a particular command when the user presses the button.

In the command design pattern one has two classes, a `Task` class that represents the command to be performed, and an `Executor` class that performs the execution of the command.

Here is the `Task` interface that we will use:

```
class Task {
public:
    virtual ~Task() {}
    virtual void execute() = 0;
};
```

Here is our `Executor` class:

```
class Executor {
public:
    /* Destructor */
    virtual ~Executor() {}
    /* Add a task to the executor */
    virtual void addTask(
        std::shared_ptr<Task> task ) = 0;
    /* Wait until all tasks are complete */
    virtual void join() = 0;
    /* Factory method */
    static std::shared_ptr<Executor> newInstance();
```

Threads   347

```
/*  Factory  method  */
static  std::shared_ptr<Executor>  newInstance(
    int  maxThreads );
};
```

You create a subclass of Task whenever you want some task to be performed. You pass that task to an Executor that decides when the task should be performed. You pass a task to an Executor by calling addTask. You can wait until all tasks are complete by calling join.

The advantage of this pattern is that it separates the decision of what tasks need to be performed from the decision of how to actually run them. For example, you could write different Executors for different architectures or performance requirements. You might, for instance, decide to write a version of Executor that waits until midnight when the computer is not so busy before executing a task.

In our example classes, the Executor returned by newInstance manages a set of threads, each of which can then perform tasks. It may be a little more efficient to reuse threads in this way than to create a new thread for every single task. This is because setting up a thread, with its associated stack and security attributes, is a relatively expensive process. It is easy to see how you could write alternative Executors that perform more sophisticated scheduling.

Notice that the programming task of implementing Task and Executor are really quite different. Writing a Task involves solving a business problem, whereas writing an Executor is more of a computer science problem. In a realistic project the tasks and the executor would be written by different people. In fact, this is going to be the case for our financial maths library too: I'll explain in detail how to implement your own Task classes, but I'll leave it to the interested reader to look at the code and work out how I've implemented Executor.

20.3   Monte Carlo pricing

As an example of how to use our scheduler, let's see how we can write a multi-threaded Monte Carlo pricer.

Monte Carlo pricing should be easy to speed up using multiple threads. Each thread can independently perform its pricing. We then just take the average result. The most tricky point is that we need to be careful about random number generation.

348   *C++ for Financial Mathematics*

20.3.1   Random number generation with multiple threads

A naive random number generator uses a global variable to store the current state of the random number generator. To use this in a multi-threaded program, we would need to add a mutex to protect this global variable. This is potentially inefficient since multiple threads will be constantly locking and unlocking the same mutex. This will mean that the cache of each processor will be frequently written to main memory.

A better approach is to use a separate random number generator for each thread. So long as we obtain appropriate seeds for the random number generators, it should be possible to ensure that the numbers generated for each thread are independent.

To implement this approach, we will want to pass around the random number generator to use whenever we call a function that generates random scenarios. You may recall that the name of the class we are using to generate random numbers is the class `mt19337`.

For this reason, the functions such as `randn` and `randuniform` in `matlib` have been updated to allow you to pass in the random number generator you would like to use:

```
/*  Create  uniformly  distributed  random  numbers  */
Matrix randuniform( int rows, int cols );
/*  Create  normally  distributed  random  numbers  */
Matrix randn( int rows, int cols );
/*  Create  uniformly  distributed  random  numbers  */
Matrix randuniform(std::mt19937& random,
                   int rows, int cols);
/*  Create  normally  distributed  random  numbers  */
Matrix randn(std::mt19937& random,
             int rows, int cols);
```

Similarly, we need to update the various `generatePricePaths` functions so that they take in a `mt19337` instance. Here is the change to the declaration of one such method on `MultiStockModel`.

```
    MarketSimulation generatePricePaths(
        std::mt19937& rng,
        double toDate,
        int nPaths,
        int nSteps) const;
```

In addition, `MultiStockModel` has a function that tells us how many random numbers it needs to generate a given stock price.

```
    /* How  many  random  numbers  are  needed
       to  generate  the  given  paths? */
    long long randSize(long long nPaths,
                       long long nSteps) {
```

Threads

349

```
        return stockNames.size()*nPaths*nSteps;
}
```

This is useful because mt19337 has a discard function that can be used to quickly update the state of the random number generator as though we had generated a given amount of random numbers but without the expense of creating that many random numbers. Using this we can launch n threads that have random number generators that will be able to generate as many independent random numbers as we need for the calculation. Note we have used a long long as the return type simply because the product might conceivably be too large to hold in an ordinary int.

20.3.2 A multi-threaded pricer

We want to modify our MonteCarloPricer so that the price method uses multiple threads. As a first step, we should rename the existing price method. Let's rename it as singleThreadedPrice

```
double singleThreadedPrice(
        int taskNumber,
        int nScenarios,
        int nSteps,
        const ContinuousTimeOption& option,
        const MultiStockModel& model ) {
```

We'll still want to call this method, it's just that our new price function will call singleThreadedPrice several times but on separate threads. It will then compute the average result across all threads.

Note that we have made another small change at the same time. The function singleThreadedPrice takes a parameter taskNumber. Together these are used to initialise the random number generator. Here is the relevant section of code from inside the implementation of singleThreadedPrice:

```
MultiStockModel subModel = model.getSubmodel(
        option.getStocks());

long long randSize = subModel.randSize(nScenarios,
                                        nSteps);
mt19937 rng;
rng.discard(randSize*taskNumber);
```

This use of discard guarantees that if we create n tasks and give them consecutive task numbers from 0–n, then the set of random numbers used for the Monte Carlo calculation will be exactly the same as if we had created a single task. In particular, all the random numbers used will be independent.

20.3.3 Implementing Task

If we wish to use our `Executor` class, we will need to write an implementation of `Task` to perform the Monte Carlo pricing. We will create a number of identical tasks, give them to the executor, and then compute the average result. The main work required by our task instances is to call `singleThreadedPrice`.

```
class PriceTask : public Task {
public:
    /*  Amount of random numbers to skip */
    int taskNumber;
    int nScenarios, nSteps;
    const ContinuousTimeOption& option;
    const MultiStockModel& model;
    /*  Output data */
    double result;

    PriceTask(
            int taskNumber,
            int nScenarios,
            int nSteps,
            const ContinuousTimeOption& option,
            const MultiStockModel& model)
        :
        taskNumber(taskNumber),
        nScenarios(nScenarios),
        nSteps(nSteps),
        option(option),
        model(model) {
    }

    void execute() {
        result = singleThreadedPrice( taskNumber,
            nScenarios, nSteps, option, model);
    }
};
```

One thing to notice about our design is that our task does not share data with other tasks except for the `const` references. Thus we are following the design recommendation given earlier to minimise the amount of communication between the threads.

Writing this `Task` class is rather tedious. It would be nice to use lambda functions instead. Doing this is left as an exercise.

20.3.4 Using the `Executor`

We now need to create a number of tasks and run them. To decide on the most appropriate number of tasks to create, we simply add a new configuration parameter `nTasks` to the `MonteCarloPricer` class.

We can now write a multi-threaded `price` function that: creates the desired number of tasks; runs them to obtain estimates for the Monte Carlo price; and combines the results to obtain a better estimate for the price.

```
double MonteCarloPricer::price(
    const ContinuousTimeOption& option,
    const MultiStockModel& model) const {
    ASSERT(nTasks >= 1);
    vector< shared_ptr<PriceTask> > tasks;
    shared_ptr<Executor> executor =
        Executor::newInstance(nTasks);
    for (int i = 0; i<nTasks; i++) {
        shared_ptr<PriceTask> task(new PriceTask(
            i, nScenarios/nTasks,
            nSteps, option, model));
        tasks.push_back(task);
        executor->addTask(task);
    }
    executor->join();
    double total = 0.0;
    for (int i = 0; i<nTasks; i++) {
        total += tasks[i]->result;
    }
    return total / nTasks;
}
```

Finally we update our tests to confirm that the result of pricing an option using multiple threads is the same result as obtained using a single thread. Since the order of calculations is slightly different there may be very tiny differences, so one does not expect precisely the same `double` value. Nevertheless, exactly the same random numbers are used in the calculation, so this should be the only source of differences.

20.3.5 Remarks upon the design

First, notice that the `Executor` class has been written to use the static factory method design pattern. We don't need to know how it was implemented in order to use it. In fact, if you look at the code in `Executor.cpp` you will see that it is very sophisticated. On the other hand, each `Task` is easy to write.

Second, notice that the only interaction with the `Task` instances happens before and after they are run.

352   *C++ for Financial Mathematics*

- We set parameters in the constructor

- We access results after calling `join` on the `Executor`

This straightforward approach is much simpler than sending data between threads using shared memory and mutual exclusion locks. Note the total separation between the algorithm to perform the financial calculation and the code that handles the threading. This is an example of how using a good design pattern can simplify your code.

20.4   Coordinating threads

20.4.1   The Pipeline pattern

The command design pattern allowed us to write a multi-threaded pricer very easily. The pattern was particularly effective because there was no communication whatsoever between the threads.

Nevertheless, sometimes you do want threads to interact with each other in more sophisticated ways. In this section we will describe one design pattern which allows two threads together in a slightly more complex manner.

Consider the following class called `Pipeline`:

```cpp
class Pipeline {
public:
    Pipeline();
    void write( double value );
    double read();
private:
    bool empty;
    double value;
    /* Mutex to coordinate threads */
    std::mutex mtx;
    /* Condition variable to signal between threads */
    std::condition_variable cv;
};
```

You can use the `Pipeline` class to connect two threads together.

Two threads share a single `Pipeline` object. One thread, the write thread, calls `write` on the `Pipeline` to store a value and then waits until the other thread has called read. Meanwhile the **read** thread calls read. The read method will wait until data becomes available to read.

The idea is that the write thread should perform part of a calculation and then store the result in the `Pipeline`. The read thread can then read the result of the partial calculation and complete the remaining processing. If

Threads   353

both threads repeat their calculations in a loop, this will result in two threads simultaneously performing the overall calculation.

As an example of a the code used to write, here is an example of a `Write-Task`:

```cpp
class WriteTask : public Task {
public:
    Pipeline& pipeline;

    void execute() {
        for (int i=0; i<100; i++) {
            pipeline.write(i);
        }
    }

    WriteTask( Pipeline& pipeline ) :
        pipeline( pipeline ) {
    }
};
```

In a similar way one can write a `ReadTask`:

```cpp
class ReadTask : public Task {
public:
    Pipeline& pipeline;
    double total;

    void execute() {
        for (int i=0; i<100; i++) {
            total+=pipeline.read();
        }
    }

    ReadTask( Pipeline& pipeline ) :
        pipeline( pipeline ),
        total(0.0 ) {
    }
};
```

Using an `Executor` we can run these two tasks simultaneously.

```cpp
static void testTwoThreads() {
    Pipeline pipeline;
    auto w = make_shared<WriteTask>( pipeline );
    auto r = make_shared<ReadTask>( pipeline );
    SPExecutor executor = Executor::newInstance(2);
    executor->addTask( r );
```

354   *C++ for Financial Mathematics*

```
executor->addTask( w );
executor->join();

ASSERT_APPROX_EQUAL( r->total, 99.0*50.0, 0.1);
}
```

Although this simple example is rather abstract, the pattern can be applied effectively. For example, consider the following code:

```
void priceByMonteCarlo() {
    double total = 0.0;
    or (int i=0; i<nScenarios; i++) {
        vector<double> path = generatePricePath();
        double payoff = computePayoff( path );
        total += payoff;
    }
    return total/nScenarios;
}
```

It looks as though this can't be parallelised using the `Command` pattern since it appears that the `generatePricePath` function is not thread safe. This is because it does not require you to pass in a random number generator as a parameter so it is certainly either not thread safe or inefficient.

Nevertheless we can parallelise this code using the `Pipeline` design. The write thread could call `generatePricePath` repeatedly and the read thread could then call `computePayoff` repeatedly on the resulting `path` objects. Of course, one would have to use a new `Pipeline` class that used a `vector<double>` to store the intermediate results rather than a simple `double`. (Note that using templates would be a good way to achieve this).

Whether this will provide a significant performance improvement is highly dependent upon how time consuming the `generatePricePath` and `compute-Payoff` methods are. What is important is that the benefit gained from using multiple processors outweighs the costs of coordinating threads. The design is particularly useful if one or more of the functions communicates across a network. Network communication is typically rather slow but not at all CPU intensive. So if you can keep your CPU busy while you're waiting to download data, this can make a big difference in making the most of your hardware.

The pipeline design is used a lot in computing. Unix users will recognise the communication betweeen processes down pipes as an example of the pipeline idea. Many banks use messaging architectures where different components perform small tasks and then forward the message to other components. This can be seen as a sophisticated form of the basic pipeline architecture.

20.4.2 How `Pipeline` is implemented

`Pipeline` is not a C++ standard class. However, it is implemented using the standard `condition_variable` class.

The `condition_variable` class is designed to enable two threads to send signals to each other about when they should start processing. If a thread finds that it has no work to do, it should call **wait** on a condition variable. This method call will not return until another thread calls **notifyAll** on the same condition variable. This basic mechanism allows one thread to pause its execution while it waits for another thread to complete a task.

The usual way of working with a `condition_variable` is to test the value of some variable to see if there is any work to do. If there is no work to do, call **wait**, otherwise perform the necessary work.

Meanwhile another thread will update the variable to indicate that there is some work to do and then call **notifyAll**. This wakes up the waiting threads which check the variable again and, observing that there is now some work to do, get on and perform the necessary work.

For this pattern to work, we will need to use a `mutex` to ensure that we synchronise the use of any shared variables `workAvailable` correctly. In general whenever you use a `condition_variable` you will have an associated `mutex`. You should hold the `unique_lock<mutex>` whenever you call **wait** or **notify_all**. You should also hold such a lock whenever you use the variables that you are using to send work between the two threads.

The full process is illustrated in Figure 20.1. In this diagram Thread 1 has to wait for Thread 2 to complete some interesting task before beginning work itself. Thread 2 communicates that work has been done by changing the value of a boolean variable called `workAvailable` from false to true. The locks and condition variables are used as described above to ensure that Thread 1 will pause its execution until there is work available and will then commence processing.

This diagram is certainly complex. In practice, the situation is not as difficult to work with as it seems. This is because one always uses condition variables in essentially the same way with the same sequence of calls. Here is a summary of how to use a `condition_variable`.

(a) Whenever you use a `condition_variable` you should also create a `mutex` to guard the data.

(b) Test if the condition is met in a while loop. You will want to lock the mutex while testing your condition.

(c) Change the data that determines whether your condition passes before calling **notifyAll**. Hold the lock while modifying the data: You should keep holding it until you have called **notify_all**.

(d) If you were to release the lock before calling **wait** it is possible that the condition may change just before you start waiting. As a result you must

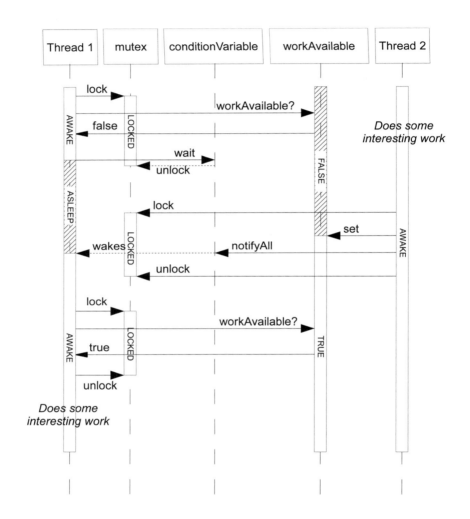

FIGURE 20.1: Sequence of calls as Thread 1 waits for Thread 2 to complete some work and set the variable workAvailable to true.

Threads 357

hold the lock using a `unique_lock<mutex>` and you must pass the lock as a parameter when calling `wait`.

(e) The `condition_variable` will release your lock and start waiting as one atomic operation.

The implementation of `Pipeline` follows this pattern exactly.

```
void Pipeline::write( double value ) {
    unique_lock<mutex> lock(mtx);
    while (!empty) {
        cv.wait(lock);
    }
    empty = false;
    this->value = value;
    cv.notify_all();
}

double Pipeline::read() {
    unique_lock<mutex> lock(mtx);
    while (empty) {
        cv.wait(lock);
    }
    empty = true;
    cv.notify_all();
    return value;
}
```

As can be seen, condition variables are much harder to understand than the `Pipeline` class itself.

The classes such as `thread` and `condition_variable` supplied as part of C++ are really designed to allow you to build threading classes that are simpler to work with, such as `Pipeline`. We emphasise again that when writing multi-threaded code you should always use established design patterns and libraries.

Exercises

20.4.1. Modify `Executor` so that you can run function objects as well as `Task` implementations.

20.4.2. Write a multi-threaded function to compute the mean of a vector using a given number of threads.

358   *C++ for Financial Mathematics*

20.4.3. Change the pipeline class so that data of an arbitrary type can be used in place of a double.

20.4.4. Write a function `integrate2d` that takes a function of two variables and integrates it over a rectangular region using the Monte Carlo method.

20.4.5. Write a multi-threaded version of `integrate2d`.

20.5   Summary

We have learned about the C++ threading library. In particular we have learned about the following C++ threading classes:

(i) `thread`,

(ii) `mutex`

(iii) `condition_variable`

(iv) `lock_guard` (used with `mutex`)

(v) `unique_lock` (used with `condition_variable`)

We have also learned some important design patterns.

- The resource acquisition is initialisation pattern.

- The command pattern

- The pipeline pattern

We have used these techniques to improve the performance of our Monte Carlo pricer. We have learned that while, in this case multi-threaded code was worth the extra effort, this is not always the case. Multi-threaded code is harder to write and test. There is a danger of *race conditions* and *deadlocks*.

Chapter 21

Next Steps

This chapter gives some suggestions of the next steps you may want to take to develop your skills as a quant.

21.1 Programming

21.1.1 Libraries

You have reached a point in your development as a C++ programmer where it no longer makes sense to write everything from scratch. It is time to learn some C++ libraries. Useful libraries for financial mathematics include:

(i) **The standard library.** We introduced some of the most useful features in Chapter 18, but it pays to be familiar with the entire library. There are excellent online resources as well as books such as [8].

(ii) **Boost.** The Boost libraries are a collection of general purpose libraries that are useful for many C++ applications. They are written to a high standard. Many of the newer standard C++ libraries were first part of the Boost library. See `http://www.boost.org`.

(iii) **Quantlib.** This is a C++ library designed specifically for financial mathematics. See `http://quantlib.org`.

(iv) **GNU Scientific Library.** The GNU scientific library contains many useful routines for numerical computations. See `http://www.gnu.org/software/gsl/`.

21.1.2 Software development

There is more to writing software than learning to program.

Your code needs to be easy to understand, easy to test, and easy to debug. Books such as [5] can enhance your approach to programming.

You need to learn how to make the most of a team of programmers and how to deliver what your customers want. [1] is a thought-provoking account of how to do this.

359

360 *C++ for Financial Mathematics*

You need to learn from the experience of other programmers. [4] describes a range of tried and tested design patterns that you can apply.

21.1.3 C++ language features

We have by no means covered the whole of the C++ language in this book.

To learn the C language features of C++, the classic reference is [9]. It can be useful to learn C as a language in its own right because many libraries such as the GNU Scientific Library are written in C. It is easier to understand their design if one knows how to program in C without C++.

We have covered the most important object-oriented features of C++, but we have not given a comprehensive overview. Instead you should consult a C++ textbook such as [16]. One issue to remember as you learn more obscure features of C++ is that a good programmer writes code that is easy to understand. This can mean choosing *not* to use certain language features because they are obscure or difficult. It is a good idea to read a book such as [13] that gives advice on which features of C++ work well and which to avoid.

We have only scratched the surface of template programming in C++. Again, a C++ textbook will tell you more about the language, but you should use a book such as [17] to learn how to use them well.

Tip: Less is more

The slogan "Less is more" was first coined by the architect Mies van der Rohe. He used it to explain his pared-down approach to architecture and design.

The slogan applies equally well in many areas of design. Most people prefer the sleek and simple interface of an iPad to the complex interface of a standard PC.

Less is more applies particularly well to programming. Using every available feature of a language can make your software harder to understand. The Java language was deliberately designed to have less features than C++ and so is easier to learn and debug. Similarly, the kernel of the Linux operating system is written in C not C++, because it is easier to reason about C programs than C++ programs.

It is a good idea to increase your knowledge of the C++ language. It is also a good idea to show restraint when you decide which features of C++ to use.

21.1.4 Other languages

If C++ is the only computer language you know, then you should learn some more languages. For many purposes C++ is not the best language to use. Other languages which are used by many financial institutions include: Python, MATLAB, R, C#, Visual Basic, and Excel. For example, if you need

Next Steps 361

to test whether a mathematical algorithm works, it will probably be quicker to prototype the code in MATLAB or Python than to write it in C++.

21.2 Financial mathematics

Our introduction to financial mathematics in Appendix A is extremely minimal. The books [14], [15], and [6] are very accessible introductions to the subject.

To keep our financial mathematics pre-requisites to a minimum, we have not covered interesting topics such as:

- pricing using trees and the related topic of PDE methods,

- pricing American options,

- pricing interest rate products,

- pricing credit products,

- improving the convergence of Monte Carlo methods,

- equity models beyond the Black–Scholes model.

Perhaps the best way to enhance your C++ skills is to learn the mathematics behind these approaches and then implement them yourself in C++. Implementing any of the above in C++ would make an excellent student project. For example, [14] gives a detailed account of how to price many different derivatives using tree-based pricing. Can you code them in C++?

It is also valuable to learn how other people have approached the problem of writing C++ libraries for financial mathematics. See for example [7] or [3].

Appendix A

Risk-Neutral Pricing

In this appendix we will briskly review the theory of risk-neutral pricing that is at the heart of much financial mathematics. The reader should not restrict themselves to this chapter if they wish to learn financial mathematics. They should begin by consulting one of the many excellent and accessible textbooks, e.g., [14], [15], and [6].

However, it will be useful to collect together the results needed in this book. In addition we will take a slightly different approach to these books. We will describe the central ideas of financial mathematics without requiring any high-powered mathematical ideas. Instead of proving theorems, we will simply refer to the numerical experiments one can perform to verify the theorems in practice. Hopefully the reader will find our presentation to be an interesting complement to more standard presentations.

A.1 The players in financial markets

To understand financial mathematics, it helps to understand who the different players are in financial markets, and what trading strategies they are using.

Probably the most obvious player in the financial markets is the *speculator*. A speculator is someone who has a theory about what will happen in the future and invests on the basis of that in order to make a profit. They take a lot of risks but hope that because of their judgement they will be able to make a large return. A famous speculator is Warren Buffett. By carefully analysing the market and the fundamentals of companies, he has successfully made billions of dollars, and in 2008 was named the richest man in the world by *Forbes*.

However, there are many other players in financial markets and many profit-making strategies. Traditionally, one important role was the *stockbroker*. They would act as a broker between different parties, some of whom wanted to buy stock and some of whom wanted to sell stock and arrange trades. With the advent of electronic trading, this role is now performed largely automatically by electronic stock exchanges. These stock exchanges make a profit simply by charging individuals to trade. The charges are quite small and often considered

363

negligible. Nevertheless, stock exchanges are important players in the financial markets. They are making money without forming any opinions on the future direction of stock prices.

Another important player in financial markets is the ordinary pensions investor. Each individual is an insignificant player, but together the pensions market is enormous. Most people do not have either the qualifications or interest to devise an investment strategy based on a careful analysis of company performance. Instead they hire someone else to manage their investments on their behalf: a pension fund manager. A pension fund manager is a member of the *financial services* industry. They make their money by performing a service for some other investor, specifically dealing with all the difficult investment questions on behalf of the pension fund members.

In actual fact, a pension fund manager will use other members of the financial services industry to help them. For example, one standard investment strategy is to "diversify" your wealth across a large number of stocks in the expectation that one's wealth will then grown in accordance with the stock market as a whole. This should be less risky than investing in individual stocks. This investment approach is based heavily on the mathematical theory of Markowitz [11]. A pension fund manager is likely to put money into investment funds that invest in large numbers of stocks and track market indices. Whoever manages these tracker funds will, of course, charge a fee for their service.

Notice how impractical it would be for most individuals to manage their pension funds without any help from the financial services industry. Following a diversification strategy where one chooses hundreds of different stocks to invest in and monitoring the performance of one's fund is a full-time job in itself. Most people are glad to pay someone else to do it for them.

In addition to ordinary individuals investing for their retirement, there are high-net-worth individuals. These are people such as dot-com billionaires and royalty who have vast amounts of money to invest but have neither the experience nor the inclination to play the stock market themselves. They too pay for financial services and often use strategies such as investing in hedge funds to manage their money. Hedge funds are simply investment funds which use more sophisticated investment strategies than basic diversification. They aim to produce investments that they hope will be appealing to their clients.

Speculators, pension investors, high-net-worth individuals, investment funds, and hedge funds are all in the business of trying to invest wisely to make a high return. They will follow quite different strategies according to how much risk they are willing to take, and their different theories about what may happen in the future. These players in the market are called the *buy side*. The name is slightly misleading because these players both buy and sell financial products. However, banks think of these market players as their customers and so think of them as the buyers of their financial services.

Tracker fund managers, on the other hand, offer a pure service. They promise that (for a cost) they will invest your money so that it tracks a particular stock market index such as the FTSE 100 or the S&P 500. Whether

Risk-Neutral Pricing

these indices go up or down, the tracker fund manager still makes money. They are purely interested in selling financial products to buy-side investors. So, tracker fund managers can be thought of as part of the *sell side*.

Banks are on the sell side. One part of their business is maintaining deposit accounts for customers and also giving loans. This part of their business is partly acting as a broker between people who want to save and people who want to borrow, and partly a matter of managing the risks of lending money. Another part of a bank's business is assisting companies with valuation, publicity, and so forth when those companies decide they want to float on the stock market. As you can see from these examples, the focus of banks is on providing a service rather than on speculation.

This book focuses on the part of a bank's business where they sell derivatives, typically to other members of the financial services industry. Derivatives are financial products that are designed to make it easier to manage risk.

One simple example of a derivative is a *call option*. A call option is a financial product that is a promise to pay at some fixed time T an amount

$$\max\{S_T - K, 0\}$$

where S_T is the price of a stock at time T and K is some pre-agreed amount called the *strike price*. If you are absolutely confident that the stock price will not fall below \$100 then you will make more money by investing in a call option with strike of \$100 than you would by investing in just the stock. However, if it turns out you are wrong and the stock price does fall below \$100, you will lose everything. So, buying a call option is a riskier strategy than investing in the stock, but potentially more rewarding. A less risk-averse investor might prefer to invest in a call option rather than in the stock itself.

Banks provide a large range of financial derivatives that allow investors to tailor their investments to very precisely match their beliefs and risk preferences. What banks wish to do is to sell such products to their customers without taking on any significant risk themselves. They are interested in making money from charges, or equivalently, by adding on a small profit margin to all the prices that they quote.

If you are going to promise to pay someone an amount equal to the value of the S&P 500 in one year's time, it is rather obvious how to hedge this risk: simply invest in the stock market in the proportions of the S&P 500. From this it is easy to work out the minimum you should charge a customer in exchange for making this promise if you want to guarantee that you don't make a loss. Note that hedging a risk simply means finding an investment strategy you can use that counteracts that risk.

It turns out that derivatives such as options can also be hedged very effectively by following certain somewhat elaborate strategies. This means that banks can sell derivatives to their customers without taking on significant risk. Customers gladly buy the derivatives because they do not have the time, expertise, inclination, or scale to follow the required hedging strategies themselves. Banks gladly take on the work, because they know that they will be

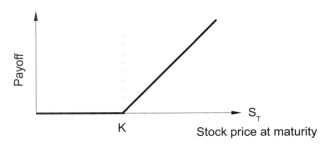

FIGURE A.1: The payoff of a call option.

able to make a profit by charging a small margin on top of the costs of the hedging strategy.

A.2 Derivatives contracts

A derivatives contract is nothing more or less than a clearly stated bet. Typically they are bets stated in terms of the behaviour of the financial markets but it is possible to buy derivatives on the weather.

A *European call option* on a stock S with *maturity* T and *strike* K is a bet on the stock price S_T at time T. If you own such an option, then at time T you will receive the following amount of money:

$$\max\{S_T - K, 0\}.$$

Figure A.1 shows the *payoff* of the call option against the value S_T.

A *European put option* on a stock S with *maturity* T and *strike* K is a bet on the stock price S_T at time T. If you own such an option, then at time T you will receive the following amount of money:

$$\max\{K - S_T, 0\}.$$

Figure A.2 shows the payoff of the put option against the value S_T.

These peculiar-seeming contracts are called options because you can express the contract in the following way. A European call option on a stock S with maturity T is the right, but not the obligation, to buy the stock at time T for the strike price K. So the customer has the *option* at time T of whether they wish to *exercise* the option and buy the stock.

To see why these formulations are equivalent, note that if at time T the stock price S_T is greater than K, the holder of the contract will certainly take the opportunity to buy the stock for a price K. This means they gain a stock worth S_T but lose K in cash. So at time T the value of their investment is

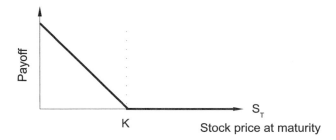

FIGURE A.2: The payoff of a put option.

$S_T - K$. On the other hand, if at time T the stock price is less than K, then it doesn't make sense to buy the stock for more than the market price. So in this case, they don't exercise the option and so their wealth remains unchanged. Taking both possibilities into account, one sees that they will end up making the amount
$$\max\{S_T - K, 0\}.$$

The word European refers to the fact that the owner of the stock is only allowed to exercise the stock at the maturity time T. When derivatives were first traded, this was the convention for option contracts in European markets. In American markets the convention was that one could choose whether or not to exercise the option at any time up to maturity. Since an American option gives you a little more choice than a European option, American option contracts will be at least as expensive as European options.

These days the typical market conventions are not segregated so much by continent. Typically, options on a single stock are American and options on an index of stocks are European. We will mostly be interested in European options in this book as they are easier to price. For this reason we will sometimes omit the word European and just talk about put and call options.

There really are no rules on what a derivatives contract *can* be. If you can invent a contract and find someone who wants to buy it off you, you have just invented a new kind of derivative. You will probably also want to think up a trading strategy for your new derivative so that you don't go bankrupt! Coming up with interesting new financial contracts is the role of a *structurer*. A structurer works within a bank to devise novel contracts that the bank believes they can hedge successfully and which the bank's customers will consider an attractive tool for managing their risks and so be willing to buy.

Here are some other derivatives contracts that we will consider in this book. It's a lot of terminology to take in, but none of the ideas are complicated.

- An Asian option with maturity T and strike K. The payoff is determined by taking
$$\max \operatorname{avg}(S) - K$$

368   *C++ for Financial Mathematics*

where avg(S) is the average stock price from time 0 up to maturity. In practice one wouldn't take the average over all times, but would take a finite average over some fixed times. Options of this form were first traded on Asian markets, hence the name.

- A Bermudan option. This is where you can choose to exercise the option at certain fixed times before maturity. It gives you more flexibility than a European option, but less than an American option. It is called Bermudan because Bermuda is halfway between Europe and America.

- A digital call option with maturity T and strike K. This has a payoff equal to 1 if the stock price at maturity is greater than K and 0 otherwise. "Binary option" or even "Boolean option" would perhaps be more logical names, but digital is the standard term. You can get European, American, etc., versions of digital options.

- A digital put option has payoff equal to 0 if the stock price at maturity is less than K and 0 otherwise.

- An up-and-out knock-out call option with maturity T and strike K and barrier B has a payoff given by:

$$\begin{cases} \max\{S_T - K, 0\} & \text{if } S_t < B \text{ for all times t } \in [0, T] \\ 0 & \text{otherwise.} \end{cases}$$

The name of this option comes from the fact that the option is worthless if the stock ever goes up to or over the barrier level B. In this case we say the option has "knocked out". The option knocks out if we ever *exceed* the barrier, which is why it is called an up-and-out option. One can similarly define an up-and-in knock-in call option by:

$$\begin{cases} \max\{S_T - K, 0\} & \text{if } S_t > B \text{ for some time t } \in [0, T] \\ 0 & \text{otherwise.} \end{cases}$$

It is reasonable to ask why anyone would want to buy any of these financial products. The general answer is that it allows people to manage their risk according to their risk preferences and their beliefs.

Let us illustrate this with an example.

Example 1: A speculator knows that a company is involved in a major court case and the court is due to make a decision in one month's time. The speculator believes that as a result, the company's stock price will probably either fall by about 15% or rise by about 15%. To exploit this theory, the speculator might buy a call option with strike equal to the current stock price plus 5% and a put option with strike equal to the current stock price minus 5%. The resulting profit will depend upon the stock price, as shown in Figure A.3. Note

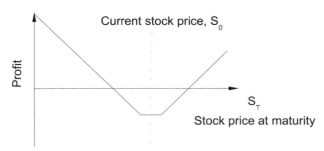

FIGURE A.3: A "strangle" consisting of a put and a call option.

that this figure shows the total profit and not just the payoff of the options because we have taken into account the cost of buying the options.

If the stock price stays the same, the speculator will lose money (the amount spent buying the options), but if the stock price either moves up sufficiently or down sufficiently, the speculator will make money. By trading in options, the speculator is able to formulate a bet that closely matches their beliefs. It is much easier to predict when a stock will move significantly than to predict whether it will go up or down, so this type of strategy is a powerful tool for a trader. This particular strategy is called a "strangle".

Notice that by buying and selling combinations of puts and calls with different strikes, it is possible to approximate pretty much any shape of payoff curve you would like. Figure A.3 is just one simple example. This explains why simple put and call options are so popular; they can be used to construct very precisely targeted bets on the future of the stock market.

Another important point is that although the word bet seems rather negative, bets can be used for insurance. Derivatives are often used for insurance.

Example 2: Consider a pension investor who is approaching retirement. It is unlikely they will want all of their wealth to be invested in the stock market as it is too risky. On the other hand, they wouldn't want all of their wealth locked up in government bonds as the interest rate is too low. One obvious solution is to invest some money in stocks and some in bonds. The problem with this approach is that it turns out you have to frequently rebalance your portfolio between stocks and bonds if you wish to invest optimally (this question was famously studied by Merton [12]). Frequently rebalancing your portfolio requires active management of your pension. In addition, rebalancing a portfolio incurs significant transaction costs. An alternative approach is to use derivatives to create a portfolio that has a risk somewhere between stocks and bonds. If you assume that puts and call options are available at all possible strikes, then it turns out that you can replicate Merton's strategy by trading in derivatives at time 0 and then simply waiting till the maturity.

370   *C++ for Financial Mathematics*

Importantly, when it comes to a practical implementation of these strategies, buying derivatives can allow you to reduce the transaction costs and largely eliminate the need for active management.

A.3   Risk-neutral pricing

Banks wish to provide the financial products desired by their customers. To make their customers happy and to ensure that there is a liquid market in their financial products, banks are typically willing to either buy or sell the same financial product. Due to the intense competition in the market, banks can only charge slightly different amounts for buying and selling the same product. Of course, they do always charge slightly different amounts for buying and selling. This difference reflects the requirement of the bank to make a profit. This willingness to both buy and sell at similar prices is perhaps the most immediately apparent difference between banks and speculators.

Banks behave very much like bookmakers in that they are willing to take either side of a bet. As a result they price their financial products in a very similar way to bookmakers.

One very interesting fact about the odds offered by bookmakers is that these odds do not necessarily reflect the chance of a particular event happening. Instead, bookmaker's odds tell you more about which bets are selling well and which are selling badly.

As an example, many people are willing to bet on their favourite football team for emotional reasons rather than because of any serious belief that they will win. They bet on what they *want* to happen rather than on what they think *will* happen. Similarly, many people don't like to bet on a firm favourite. This is because even if their horse wins they might not make enough money to make it exciting. Most people bet for entertainment as much as profit. A 1% return on an investment in a day is, from a financial markets point of view, an excellent rate of return. However, if you are at a race track and bet £100 on the only four-legged horse running a particular race in the hope of making £1 profit, then you probably won't find your watching the race to be particularly exciting.

As a result, bookmakers do not trouble themselves with carefully studying the physiques and pedigrees of horses when they choose what odds to offer. Instead they simply match up the bets between different parties in such a way that they are guaranteed to make a profit. This profit comes from the small difference in the price offered by bookmakers when taking the different sides in any particular bet.

The derivatives desks of banks operate in almost exactly the same way.

Risk-Neutral Pricing

Derivatives traders are largely indifferent to whether a stock goes up or down in much the same way as bookmakers are indifferent as to which horse wins a race.

Since the margin between the buy and sell prices of both bookmakers and banks are rather small, they find it a useful approximation to think of themselves as offering a single price at which they will either buy or sell. In practice, both banks and bookmakers can first choose a self-consistent set of prices at which they are willing to either buy or sell. Then when a customer asks to buy a particular product, they quote a slightly higher price, and a slightly lower price when their customer wishes to sell. In this way, they expect to make a small profit from each event that their customers bet on.

It is vital that the prices offered by the bank or the bookmaker are self-consistent in some sense. What they don't want to happen is for a customer to be able to place a complex bet that the customer knows they are guaranteed to win. Such a bet is called an arbitrage.

To ensure that such a bet is impossible, bookmakers price their bets *as if* the cost of a bet was determined by the expected payoff in some probability model. As one knows, the odds offered by bookmakers *look* as though they are derived from the probability of each horse winning. Yet as we have discussed, those prices are derived by considering the market for bets and not by considering the real chance of those events occurring.

Derivatives traders work in exactly the same way. One slight difference is that derivatives traders offer bets over longer time periods than bookmakers, so they factor charging and paying interest into their calculations.

The probability model used by bookmakers and derivatives traders is called a $\mathbb{Q}$-measure model. This is a model that looks exactly like an ordinary probability model, but the values it computes are measures of price and not measures of the chance of an event happening. A classical probability model that attempts to measure the chance of an event taking place is called a $\mathbb{P}$-measure model.

If the interest rate for a risk-free bank account is at a constant level r so that a principal P invested is worth Pe^{rt} at time t, then the amount charged by a derivatives trader for an option that pays $f(S_T)$ at time T is given by the following formula:

$$\text{Risk neutral price} := e^{-rT} \, \mathrm{E}_{\mathbb{Q}}(f(S_T)).$$

Here $\mathrm{E}_{\mathbb{Q}}$ denotes the expectation taken using the $\mathbb{Q}$-measure model. The factor e^{-rT} is a discount factor to take account of interest payments. The motivation for calling this price the "risk neutral price" will be explained shortly.

Notice that if you invest an intial sum P in the risk-free bank account, then the payoff will be Pe^{rT} and is independent of the stock price. Hence, its expectation in the $\mathbb{Q}$-measure model will also be Pe^{rT}. So the "risk neutral" valuation of a bank account in which P is invested at time 0 is:

$$e^{-rT} \, \mathrm{E}_{\mathbb{Q}}(Pe^{rT}) = P.$$

372   *C++ for Financial Mathematics*

This provides a basic test of the risk-neutral valuation formula's consistency: It can correctly value risk free investments.

Kolmogorov's axioms

For a pure mathematician, intuitive notions such as "chance" have no meaning. To do rigorous mathematics, one needs axioms for probability theory from which to rigorously deduce results. Kolmogorov's axioms provide this basic foundation. One advantage of the mathematical approach to problem solving is that the same mathematical model can often be applied to different phenomena. For example, the theories of gravity and electricity both obey inverse square laws and so can be modelled in the same way. It is useful to think of the mathematical theory of probability as a purely abstract theory which can then be applied to either model chance events (using probability as a $\mathbb{P}$-measure) or model prices (using probability as a $\mathbb{Q}$-measure). Probability theory can also be applied to study subjects which appear to have nothing to do with chance. For example, probability theory can be used to study integration.

The mathematical details of Kolmogorov's axioms can be found in any rigorous introductory book on probability. Speaking loosely, a probability model is defined to be a function from a set of events to the interval $[0, 1]$, which assigns the "probability" to each event. Note that mathematically, this model is simply a function. Interpreting this function in terms of chance or in terms of pricing is outside the scope of strict mathematics. This function can be seen as a measure of the "size" of a set of events, with larger sets being those with a higher probability of occurring. The mathematical theory which makes this precise is called measure theory, and Kolmogorov's axioms are stated in terms of measure theory. This is where the word measure comes from when we discuss the $\mathbb{P}$-measure and the $\mathbb{Q}$-measure.

A.4   Modelling stock prices

What would be a good $\mathbb{Q}$-measure probability model for stock prices and what would be a good $\mathbb{P}$-measure model?

One idea is to assume that the stock price moves up and down by a normally distributed amount at each time step. If we choose a time step δt we can model a stock price by assuming that it starts at a given level S_0 at time 0, and then model the values at subsequent times δt, $2\delta t$, $3\delta t$, and so forth using the equation:

$$S_{t+\delta t} = S_t + X_t$$

Risk-Neutral Pricing 373

where the random variables X_t are independent and normally distributed, with some fixed mean and standard deviation. This model is called the discrete *Bachelier model* after Louis Bachelier, one of the pioneers of financial mathematics. (It is a discrete model because it only models the stock price at discrete time steps δt.)

The problem with the Bachelier model is that if the stock price is \$10 then \$1 would be quite a big change in the stock price. Whereas if the stock price is \$1000, then \$1 would be a relatively small change in the stock price. The Bachelier model assumes the size of the increments are independent of the current stock price and so it doesn't have the correct scaling behaviour. To fix this, we simply assume that the log of the stock price has independent normally distributed increments with a fixed mean and standard deviation. So we write

$$\log(S_{t+\delta t}) = \log(S_t) + Y_t \tag{A.1}$$

where the Y_t are independent, normally distributed random variables with some fixed mean and standard deviation.

We can now write each Y_t in terms of some other normally distributed random variables ϵ_t which have mean 0 and standard deviation 1. If A is the mean of each Y_t and B is the standard deviation, we have

$$Y_t = A + B\epsilon_t.$$

Let us suppose there are $N = \frac{1}{\delta t}$ time steps from 0 to 1. Using the formulae for the expectation and variance of a sum of random variables, we can compute the mean and variance of $\log S_t$ as follows.

$$E(\log(S_1) - \log(S_0)) = NE(Y_t) = \frac{A}{\delta t}.$$

$$\mathrm{Var}(\log(S_1) - \log(S_0)) = N\,\mathrm{Var}(Y_t) = \frac{B^2}{\delta t}.$$

In finance we use units of one year to measure times. So the first of these formulae says that the log of the stock price grows by $\frac{A}{\delta t}$ on average each year. The second formula can be interpreted similarly in terms of the standard deviation over a year.

Let us write $\tilde{\mu}$ for the expected growth in the log of the stock price over a year and σ for the variance in the growth of the log of the stock price over a year.

We can then compute A and B in terms of $\tilde{\mu}$ and σ. We have:

$$A = \tilde{\mu}\delta t$$
$$B = \sigma\sqrt{\delta t}.$$

This allows us to rewrite our model (A.1) as follows:

$$\log(S_{t+\delta t}) = \log(S_t) + \tilde{\mu}\delta t + \sigma\sqrt{\delta t}\epsilon_t. \tag{A.2}$$

374 *C++ for Financial Mathematics*

The advantage of reformulating the model in this way is that we have a clear statement of how the constants $\tilde{\mu}$ and σ can be interpreted in terms of the change of the log of stock price over a year. By contrast, the choices for A and B depend upon the choice of time interval δt and so don't have such a natural interpretation.

The sum of independent normally distributed random variables is also a normally distributed random variable, so in our model, the distribution for the log of the stock price is normal at all times. We can compute the mean and variance at any subsequent time T just as we computed the mean and variance at time 1 above. We conclude that at any time t, according to this model, the log stock price is normally distributed with mean $\log S_0 + \tilde{\mu}t$ and standard deviation $\sigma\sqrt{t}$. We write:

$$\log S_t \sim N(\log S_0 + \tilde{\mu}t, \sigma\sqrt{t}).$$

This means that we can also compute the distribution of S_t at each time. Since $\log S_t$ is normally distributed, we say that the S_t is log-normally distributed.

Let us compute the probability density function of S_t according to this model. We can write down the probability density of $z_t := \log S_t$. It is

$$p_{z_t}(x) = \frac{1}{\sigma\sqrt{2\pi t}} \exp\left(-\frac{(x - \log(S_0) - \tilde{\mu}t)^2}{2\sigma^2 t}\right). \tag{A.3}$$

We can use this to compute the expected value of $S_t = \exp(z_t)$. The necessary integration is left as an exercise. The result is

$$E(S_t) = S_0 \exp\left(\left(\tilde{\mu} + \frac{1}{2}\sigma^2\right)t\right).$$

Because the right-hand side is the same as the growth of a principal S_0 invested at a continuously compounded interest rate of $\tilde{\mu} + \frac{1}{2}\sigma^2$, it is natural to introduce a variable $\mu := \tilde{\mu} + \frac{1}{2}\sigma^2$ called the *drift*. Since investors already understand the notion of interest rates very well, the drift is a somewhat more intuitive notion to an experienced investor than $\tilde{\mu}$. So let us rewrite our model in terms of μ.

Definition. Discrete-time geometric Brownian motion

$$\log(S_{t+\delta t}) = \log(S_t) + \left(\mu - \frac{1}{2}\sigma^2\right)\delta t + \sigma(\delta t)^{\frac{1}{2}}\epsilon_t. \tag{A.4}$$

We will call this model the discrete-time geometric Brownian motion model. Since in this book we do not make much use of the mathematical theory of continuous-time stochastic processes, we will some times just call this model simply geometric Brownian motion. The parameter μ is the drift and can be interpreted as a form of interest rate as discussed above. The parameter σ is called the volatility. It measures the size of random fluctuations in the stock market.

Risk-Neutral Pricing 375

In [2], Black and Scholes proposed that the continuous-time version of this model could be used as a $\mathbb{P}$-measure model for stock prices. For this reason we will sometimes refer to this model as the Black–Scholes model for stock prices.

To answer whether it is a good $\mathbb{P}$-measure model, one would need to perform statistical tests to see how well the model fits real-world data. The data suggests that this is only a rather crude approximation of the real behaviour of stock prices. One notable gap in the model is that it underestimates the probability of large changes in the stock price. Another notable gap in the model is that the volatility is constant, but it appears to change over time. For real financial applications, therefore, the discrete-time geometric Brownian motion might not be a good choice of $\mathbb{P}$-measure model for stock prices. However, for teaching purposes it is a good choice because it is relatively easy to write down and analyse.

Is discrete-time geometric Brownian motion a good choice for a $\mathbb{Q}$-measure model of stock prices? In general it is not, because it would give inconsistent price information. Recall that in a $\mathbb{Q}$ measure model the price of financial products is given by the discounted expected value of the financial product. We can use this to calculate that the stock price at time 0 is given by:

$$S_0 \exp((\mu - r)t).$$

On the other hand, by definition the stock price at time 0 is S_0. Therefore discrete-time geometric Brownian motion can only be a $\mathbb{Q}$ measure model if we have $\mu = r$ where r is the risk-free interest rate. We make the following definition.

Definition. *The $\mathbb{Q}$-measure geometric Brownian motion model for the stock price is given by*

$$\log\left(\frac{S_{t+\delta t}}{S_t}\right) = \left(r - \frac{1}{2}\sigma^2\right)\delta t + \sigma(\delta t)^{\frac{1}{2}}\epsilon_t. \tag{A.5}$$

It is important to notice that most people would agree immediately that this is not a plausible $\mathbb{P}$-measure model. The motivation of most investors for investing in the stock market is that they expect a better rate of return from stocks than from risk-free investments. It follows immediately for these investors that their $\mathbb{P}$-measure model cannot be a $\mathbb{Q}$-measure model. In the case of the model (A.4) one would expect that a $\mathbb{P}$-measure model would have $\mu > r$. This is where the phrase *risk neutral pricing* comes from. The $\mathbb{P}$-measure and the $\mathbb{Q}$-measure only coincide for highly unusual investors who are indifferent as to whether their investments are risky. Such investors are called *risk-neutral* investors. Typical pensions investors are *risk averse*. Gambling addicts are *risk seeking*.

While we have seen that Equation (A.5) is not a good $\mathbb{P}$-measure model for anyone but risk neutral investors, is it a good choice of $\mathbb{Q}$-measure model? It is certainly a probability model that gives the correct initial value of the

376 C++ for Financial Mathematics

stock price when one can make risk-free investments at a rate of r. This means that a trader who prices derivatives using this model cannot be arbitraged. However, this is only a minimum requirement for a $\mathbb{Q}$-measure model. It is also important to check if it fits market prices for derivatives well. To check this we should compute the risk-neutral prices of put and call options. One simply needs to compute the expected value of the payoff

$$\max(\exp(z_t) - K, 0)$$

using the formula for $p(z_t)$ given in Equation (A.3). This is another exercise in integration.

Theorem 1. The Black–Scholes Formulae *If one uses the $\mathbb{Q}$-measure geometric Brownian motion model* (A.5) *to price stock options, then the price of a European call option at time* 0 *with maturity T and strike K is*

$$C = N(d_1)S_0 - N(d_2)K \exp(-rT) \tag{A.6}$$

where N is the cumulative distribution function of the standard normal distribution,

$$d_1 = \frac{1}{\sigma\sqrt{T}} \left(\log\left(\frac{S}{K}\right) + \left(r + \frac{\sigma^2}{2}\right)\sqrt{T} \right),$$

and

$$d_2 = \frac{1}{\sigma\sqrt{T}} \left(\log\left(\frac{S}{K}\right) + \left(r - \frac{\sigma^2}{2}\right)\sqrt{T} \right).$$

The corresponding formula for a European put option is:

$$P = N(-d_2)K \exp(-rT) - N(-d_1)S. \tag{A.7}$$

To see if our $\mathbb{Q}$-measure model accurately fits market data, we should see if we can find parameter values so that the prices given by the formulae above match real market prices for options. This is left as a rather tricky exercise for you to try out using the C++ skills developed in this book (Exercise 19.4.5). The answer is that our $\mathbb{Q}$ model gives a basic crude fit to the market data for options, but a more sophisticated model is needed in practice.

We should emphasise that Black and Scholes used a far more elaborate argument to derive these formulae based on $\mathbb{P}$-measure models in their seminal paper [2]. We have simply assumed that the pricing model takes a particular form and then computed some integrals. We will discuss Black and Scholes' original argument in more detail when we discuss hedging.

Tip: Dimensional analysis

It is conventional to use the year as the unit of time in finance calculations. Often the units are omitted when writing financial formulae. For example, one might talk about an interest rate of 5% when one perhaps should state more clearly that it is an interest rate of 5% per annum. Thus, the dimensions of an interest rate are T^{-1} where T stands for the time dimension. The dimensions of the drift parameter μ are also T^{-1}. One can see from the above formulae that the dimensions of the volatility parameter σ must be $T^{-\frac{1}{2}}$. This states that the standard deviation of the change in the log of the stock price grows at a rate proportional to the square root of time. This is simply a consequence of the formula for computing the variance of a number of independent increments. Volatilities are often quoted as a percentage, so a volatility of 20% means that $\sigma = 0.2$.

You can use this dimensional analysis to check various formulae in financial mathematics. For example, we can rewrite the discrete-time Black–Scholes model (A.4) as

$$\log\left(\frac{S_{t+\delta t}}{S_t}\right) = \left(\mu - \frac{1}{2}\sigma^2\right)\delta t + \sigma(\delta t)^{\frac{1}{2}}\epsilon_t. \tag{A.8}$$

In this formula the stock price has units of (say) dollars, but these units cancel in the computation of $\frac{S_{t+\delta t}}{S_t}$ to give us a dimensionless quantity. Similarly, all the terms on the right-hand side are dimensionless.

If you have heard of *fractals* and the associated theory of fractional dimensions, you may be interested to notice that the fractional dimension of volatility is connected to the fact that stock prices are fractals.

A.5 Monte Carlo pricing

To compute risk-neutral prices of more complex derivatives, one can use the algorithm of Monte Carlo pricing. The algorithm is as follows.

- *Simulate N scenarios for the asset prices using your $\mathbb{Q}$-measure model.* A scenario is simply one possible future set of asset prices. These asset prices only need to be computed at the times relevant to the derivatives contract. For a European option this means you only need to compute prices at maturity. For a discrete-time knock-out option, you would need to compute prices at intermediate times to see if the option has knocked out. Your simulation should have the same probability distribution as your $\mathbb{Q}$-measure model.

- Compute the mean final payoff of the derivative.

- Discount the mean payoff to take account of the interest rate. This is an estimate for the price of the derivative.

The mathematical theorem underpinning this approach is the central limit theorem. It tells us that the expected price computed in this way is equal to the risk-neutral price. Moreover, it tells us that the standard error in this estimate is proportional to $\frac{1}{\sqrt{N}}$. So by increasing the number of scenarios in the simulation one can get increasingly accurate estimates of the derivative price. We can also compute confidence bounds for the error using the central limit theorem.

Monte Carlo pricing is not necessary if the payoff of the derivative depends upon a single stock price at a single time. Recall that the risk-neutral price is defined as an expectation which is, in turn, defined as an integral. If the price of the derivative depends on the stock price at a single time, one only needs to calculate a 1-dimensional integral. This can be done more efficiently using Simpson's rule than by using the Monte Carlo algorithm.

If the payoff of the derivative depends upon the stock price at multiple time points or upon multiple risk factors, then the relevant integral will be a higher-dimensional integral. In general, high-dimensional integrals are difficult to perform efficiently and some form of Monte Carlo algorithm is required. The Monte Carlo algorithm we have given above for pricing derivatives is just an application of the Monte Carlo integration formula.

Let us state the result explicitly even though it is simply a special case of the central limit theorem.

Theorem 2. *Let $f : [0,1]^d \to \mathbb{R}$ be an integrable and square integrable function. Then one can estimate the d-dimensional integral $\int f$ by*

$$\frac{1}{N} \sum_{j=1}^{N} f((x_1^j, x_2^j, \ldots, x_d^j))$$

where the x_i^j are uniformly distributed random variables on the interval $[0,1]$. By this we mean that the sum on the right hand side has expectation equal to $\int f$ and

$$\sqrt{N} \left(\frac{1}{N} \left(\sum_{j=1}^{N} f((x_1^j, x_2^j, \ldots, x_d^j)) \right) - \int f \right)$$

converges in distribution to a normal distribution $N(0, \sigma^2)$.

It is important to know that section has simply described the most basic form of Monte Carlo pricing. Many refinements are possible. For a simple example, see Exercise 9.3.7. One should consult a book dedicated to Monte Carlo pricing to understand the state of the art. Implementing more sophisticated Monte Carlo algorithms would be an excellent test of the C++ skills you will develop by reading this book.

We remark that Monte Carlo techniques can be used for many purposes other than just calculating risk-neutral prices. Monte Carlo methods can also

Risk-Neutral Pricing

be used for calculating risk figures, developing optimal trading strategies, and testing the effectiveness of trading strategies.

We also note that Monte Carlo methods are not the only way of pricing derivatives. We will see in the exercises that simple derivatives can be priced using familiar integration rules such as the rectangle rule or Simpson's rule. There are also pricing methods based around trees and partial differential equations.

A.6 Hedging

We have described how a trader can produce prices that are guaranteed to be arbitrage-free if they use the risk-neutral pricing methodology. However, this is by no means enough to guarantee that a trader who charges these prices will make money.

This is obvious if one considers bookmakers. Bookmakers need to produce a consistent set of prices at which they are willing to buy and sell different bets. However, that is not all they must do. They need to work out how to *hedge* their bets. This is the process of balancing the bets of different customers off against each other in such a way that the bookmaker can be sure that they will not lose money. Note that another way of hedging bets is to place your own counter-bets with another bookmaker, so it is not necessary to wait for appropriate customers to arrive before you can start hedging bets.

In general you will need to place at least one hedge for every source of risk that you are potentially exposed to. In a simple model of a stock market with a single stock, following the geometric Brownian motion (A.4) and with a fixed risk-free interest rate, the only source of risk is the stock price. Thus, a trader who sells derivatives will want to place at least one bet to reduce their risk to changes in the stock price.

One way to measure the amount of risk of your portfolio to changes in the stock price is to compute the sensitivity to changes in the stock price. If P is the value of your portfolio, the sensitivity:

$$\frac{\partial P}{\partial S}$$

gives a measure of how much your portfolio would change in value if the stock price changed, everything else being fixed. This sensitivity is called the delta of the portfolio and is denoted by Δ.

Clearly the delta of a portfolio consisting solely of n units of stock is simply n and the delta of a portfolio consisting of only risk-free investments is 0. This means that by borrowing money to sell stock or by selling stock and placing the money in risk-free investments, one can change the delta of one's portfolio. We assume that any amount of stock can be bought or sold at a given price

380   *C++ for Financial Mathematics*

and similarly that any amount can be saved in the risk-free bank account, and that any amount can be borrowed at the same rate. This means that it is possible to adjust the delta of a portfolio to any desired value simply by transferring money between the risk-free account and stock.

The assumption that one can buy and sell any desired amount of stock and that one can lend or borrow unlimited amounts all at the same rate is a simplifying assumption. The essence of this assumption is that one is a small player in a large market and that the costs of trading are negligible. Notice that we allow one to sell stock one does not own in order to obtain a portfolio containing a negative quantity of stock. This is just a mathematical shorthand for borrowing stock off someone else in the same way that a negative bank account is shorthand for being in debt.

Under this assumption, it is always possible to trade in stock to ensure that the delta of your portfolio is 0. This means that, to first order, the value of your portfolio will not change when the stock price changes. This is as good a hedge as it is possible to make at any one time. The *delta hedging trading strategy* is to regularly trade in stock in order to ensure that the delta of one's portfolio is always close to zero. This trading strategy is the key to the success of risk-neutral pricing.

A trader who pursues this strategy charges the risk-neutral price for derivative products and then follows the delta hedging trading strategy. The idea is that since the sensitivity to changes in the stock price is kept at 0, this strategy should be close to risk free (at least if one measures risk using the $\mathbb{Q}$ probabilities). The effectiveness of this strategy depends upon how frequently one trades to keep the delta close to zero. In the limit as one approaches delta hedging in continuous time, the risk of the strategy actually drops to zero. We will not attempt to prove this mathematically. Instead, in Chapter 14 we confirm this theory numerically by plotting a histogram of the profit and loss of this investment strategy and seeing how it changes as the time interval decreases. Notice also that the method can be generalised to multiple risk factors, to hedge one needs to invest in financial products in such a way that the overall sensitivity to each risk factor is zero.

The obvious objection to this strategy is that it measures the risk of investments using $\mathbb{Q}$ probabilities. However, if one believes that the $\mathbb{Q}$ probabilities that reflect market prices aren't so different from the actual $\mathbb{P}$ probabilities of events occurring, then one might hope that the risk of the strategy will still be low even when we measure the risk using $\mathbb{P}$ probabilities.

This raises the question of how the delta hedging strategy performs if one delta hedges using a given set of $\mathbb{Q}$ probabilities, but stock prices actually evolve according to some other set of $\mathbb{P}$ probabilities. One approach to answering this question is numerical experiments. We pursue this approach in Chapter 14. Indeed, one can perform quite sophisticated tests of how the delta hedging strategy performs under a variety of constraints. For example, one can study what happens if one pursues the delta hedging strategy derived using the $\mathbb{Q}$ model (A.5) but the stock price actual evolves using a completely

Risk-Neutral Pricing 381

different model, perhaps including features such as fat tails and transaction costs. These numerical results are enough to explain whether or not trading derivatives and investing them using the delta hedging strategy is likely to be a profitable strategy. These results can also indicate how much one should charge in order to make an acceptable profit even in the face of issues such as transaction costs.

Simple though these numerical experiments are to perform, this does not reflect the history of the development of the theory of risk neutral pricing. In [2], it was proved mathematically that if the stock price follows the $\mathbb{P}$-measure model (A.4), then the delta hedging strategy in continuous time is risk-free so long as one uses the $\mathbb{Q}$-measure model (A.5). This has the consequence that if you believe that the continuous-time version of the $\mathbb{P}$-measure model (A.4) is a perfect model for stock prices, then the only possible consistent way of pricing derivatives is to use the $\mathbb{Q}$-measure model (A.5).

This mathematical result has had a profound impact on financial markets. It provides a form of mathematical explanation of how a bank can trade in derivatives to make a profit.

However, the mathematical result requires a number of unrealistic assumptions. In particular:

- Unlimited amounts of stock can be bought or sold at this price and one is not restricted to buying whole numbers of stocks (we say there is infinite liquidity).

- There are no transaction costs.

- It is possible to trade in continuous time.

- The stock price follows the continuous-time limit of the model (A.5).

- Any amount of money may be invested or borrowed at a fixed risk-free rate of r.

- The stock does not pay a dividend.

More sophisticated modelling can be used to remedy the last three problems and still obtain similar mathematical results, but the first three assumptions are essential to the argument.

In some ways, therefore, the numerical results of Chapter 14 are more convincing than the mathematical result as one can easily test the performance of the strategy with more realistic market models.

It is clear empirically that assumptions such as no transaction costs, infinite liquidity, constant volatility, and log normally distributed stock prices are unrealistic. More subtly, the assumption that the trader starts with their own view on the likely evolution of the market (i.e., a fixed $\mathbb{P}$-measure model) is also typically false. Just as bookmakers have little or no opinion about the racing-form of horses, so derivatives traders do not need to develop a careful opinion about the stock market. In practice, derivatives traders choose a basic form

382   *C++ for Financial Mathematics*

for a $\mathbb{Q}$-measure model and then fit the parameters to actual market data for put and call prices. This process is called *calibration*. The traders then simply assume that the difference between the $\mathbb{P}$-measure model and their pricing model is sufficiently small that their hedging strategy and pricing policy will still turn an acceptable profit with an acceptable risk.

While the mathematical derivation of the Black–Scholes model and the delta hedging strategy depend heavily upon a number of implausible assumptions, the practical strategy of delta hedging works effectively in practice so long as one chooses good $\mathbb{Q}$-measure models. The most convincing proof of this is the financial success of the derivatives desk of banks.

It is important to notice that the risk-neutral pricing approach together with delta hedging is just one possible investment strategy.

An alternative mathematical approach is to price financial products using the notion of expected utility. A utility function assigns a number to the future wealth of an investor that measures how happy that investor is with the outcome. One can then model the behaviour of speculators by assuming that they are trying to maximize their expected utility in a $\mathbb{P}$ model that matches the beliefs of that investor. Using this model, speculators attempt to derive their profits by forming more accurate beliefs than other people and hence out-performing the market. This is an entirely different strategy to the approach of simply following the market and making a profit by charging a little more to buyers than one pays to sellers. In particular, speculators are willing to take large risks in the hope of making large profits. Delta hedgers try to avoid risks entirely and instead aim to make a profit by offering a valuable service.

A.7   Summary

We have described the delta hedging strategy used by banks. This strategy allows banks to make a profit by, in effect, selling a risk-management service to their customers. Risk-neutral pricing is a key component of the delta hedging strategy. Risk-neutral pricing tells you what you should charge your customers if you are following the delta hedging strategy. In addition, the partial derivatives of the risk neutral price tell you how to hedge your risk when you are following the delta hedging strategy.

Other trading strategies, and hence other pricing methodologies, are used by other types of investors.

Bibliography

[1] K. Beck. *Extreme Programming Explained: Embrace Change*. Addison-Wesley Professional, 2000.

[2] F. Black and M. Scholes. The pricing of options and corporate liabilities. *The Journal of Political Economy*, pages 637–654, 1973.

[3] D. J. Duffy. *Financial Instrument Pricing Using C++*. John Wiley & Sons, 2013.

[4] E. Gamma, R. Helm, R. Johnson, and J. Vlissides. *Design Patterns: Elements of Reusable Object-Oriented Software*. Pearson Education India, 1995.

[5] A. Hunt and D. Thomas. *The Pragmatic Programmer: from Journeyman to Master*. Addison-Wesley Professional, 2000.

[6] M. S. Joshi. *The Concepts and Practice of Mathematical Finance*, volume 1. Cambridge University Press, 2003.

[7] M. S. Joshi. *C++ Design Patterns and Derivatives Pricing*, volume 2. Cambridge University Press, 2008.

[8] N. M. Josuttis. *The C++ Standard Library: a Tutorial and Reference*. Addison-Wesley, 2012.

[9] B. W. Kernighan, D. M. Ritchie, and P. Ejeklint. *The C Programming Language*, volume 2. Prentice-Hall, 1988.

[10] W. Margrabe. The value of an option to exchange one asset for another. *The Journal of Finance*, 33(1):177–186, 1978.

[11] H. Markowitz. Portfolio selection. *The Journal of Finance*, 7(1):77–91, 1952.

[12] R. C. Merton. Lifetime portfolio selection under uncertainty: The continuous-time case. *The Review of Economics and Statistics*, pages 247–257, 1969.

[13] S. Meyers. *Effective C++: 55 Specific Ways to Improve Your Programs and Designs*. Pearson Education, 2005.

384   *Bibliography*

[14] S. E. Shreve. *Stochastic Calculus for Finance I: The Binomial Asset Pricing Model.* Springer Science & Business Media, 2004.

[15] S. E. Shreve. *Stochastic Calculus for Finance II: Continuous-Time Models*, volume 11. Springer Science & Business Media, 2004.

[16] B. Stroustrup. *Programming: Principles and Practice Using C++.* Pearson Education, 2014.

[17] D. Vandevoorde and D. M. Josuttis. *C++ Templates.* Addison-Wesley Longman Publishing Co., Inc., 2002.

Index

/*...*/, multiple line comment, 8
//, single line comment, 8
&, bitwise and, 29
&, obtaining a memory address, 181
&, references, 101
&&, logical and, 29
->, pointer operator, 182
\a, bell character, 22
\n, newline character, 9
\r, carriage return character, 22
\t, tab character, 22
^, exclusive or, 29
Δ stock price delta, 379
., dot operator, 126, 129
=, assignment operator, 29
=0, abstract method, 162
==, comparison operator, 29
|, bitwise or, 29
||, logical or, 29
!=, not equals operator, 29
? :, ternary operator, 66
*, pointer operator, 181
*, type modifier, 182
~, not operator, 29

abstract functions, 212
American option, 367
antithetic sampling, 157
arbitrage, 371
array programming, 286
arrays, 176
ASCII, 22
Asian option, 368
ask price, 243
assembly language, 3
ASSERT, 87
assertions, 87

assignment operator, 282
assignment operators, 30
auto, 306

Bachelier model, 373
base class, 207
Bermudan option, 368
bid ask spread, 243
bid price, 243
bit, 18
Black–Scholes formula, 376
Black–Scholes model, 375
boiler plate, 8
bool, 20
Boolean, 20
break, 60
breakpoint, 250, 256
bug tracking, 259
build, 260
byte, 18

call option, 365, 366
capturing variables, 332
casting, 22
catch, 63
Central Limit Theorem, 378
char, 20
char*, C style string, 27, 185
character, 20
child class, 209
Cholesky decomposition, 292, 321
cin, 9
class, 124
class keyword, 125
code-review, 246
comma operator, 67
command pattern, 346

385

comments, 8
compiled language, 3
compiler, 3
concurrent programming, 338
condition_variable, 355
console, 39
const, 47, 102, 130, 188, 274
constructor, 138
continue, 60
continuous integration, 261
copy constructor, 283
cout, 9
CPU, 17

database, 345
deadlock, 343
DEBUG_PRINT, 88
debugger, 247
declaration, 42, 125, 296
default, 63
default constructor, 139
definition, 42, 80, 130, 296
delete, 192
delete [], 179
delta, 157, 379
delta hedging, 233, 380
derivatives, 366
destructor, 269
development environment, 1
digital option, 368
do, 57
double, 9, 20
down-and-out option, 368

else, 13
encapsulation, 135
error handling, 61
escaping, 22, 115
European option, 366
extends, 208
extreme programming, 91

factory design pattern, 227
false, 20
float, 9, 20
for, 58, 309

forward declaration, 217
function, 37
function object, 329
function pointers, 333
functor, 329

general protection fault, 203
geometric Brownian motion, 375
global variable, 47
GPU, 287

header file, 72, 77
heap, 200, 202
hedging, 379
hexadecimal, 18
HTML, 117

if, 13, 32
include, 28, 72, 77
indifference pricing, 382
INFO, 88
information hiding, 78
inheritance, 207
initialisation list, 140
initializer_list, 315
inline, 80
inlining member functions, 205
instance, 124
int, 9, 18
interface, 159
interpreted language, 3
iterators, 307

kilobyte, 18
knock-in option, 368
knock-out option, 368
Kolmogorov's axioms, 372

lambda function, 330
libraries, 359
liquidity, 381
list, 312
local classes, 171
local variable, 47
logging, 246, 261
long, 19

Index

machine code, 3
macros, 86
`main`, 39
`Makefile`, 6
`make_shared`, 196
`map`, 315
Margrabe option, 326
`Matrix`, 267
maturity, 366
measure theory, 372
megabyte, 18
member function, 129
member variable, 125
memory address, 26
memory leak, 203, 262
Mersenne Twister, 154
Merton problem, 370
messaging architectures, 354
method, 124
modularity, 40
Monte Carlo pricing, 145, 377
move operator, 285
`mt19337`, 154
multiple inheritance, 215
`mutex`, 340
mutual exclusion, 339

`namespace`, 48
`new`, 192
`new []`, 179
nibble, 18
null pointer exception, 203
null-terminated string, 185
`nullptr`, 194

object, 124
`ofstream`, 104
once and only once, 40
operator overloading, 275
operators, 28
`ostream`, 104
overloading functions, 46
`override`, 211

pair programming, 246
parameter, 38

parent class, 209
pass by reference, 101
pass by value, 101
pipeline, 352
$\mathbb{P}$-measure, 371
pointer, 27, 179
pointer arithmetic, 182
polymorphism, 166
`#pragma once`, 77
`private`, 135
private inheritance, 215
process, 337
profiling, 262
`protected`, 220
pseudo square root, 321
`public`, 125
put option, 366

$\mathbb{Q}$-measure, 371

race condition, 340
random numbers, 154
recursion, 41
references, 101
release, 12
resource acquisition is initialisation
 (RAII), 341
`return`, 38, 60
return by reference, 280
risk-neutral pricing, 370
rule of three, 282

scope, 47, 65
seed, 155
SEGV, 203
`set`, 310
`shared_ptr`, 194
`short`, 19
short circuit evaluation, 66
`sizeof`, 28
`size_t`, 20
smart pointer, 194
stack, 200
stack trace, 248, 253
standard template library (STL), 303
`static`, 79, 218

388   *Index*

static analysis, 261
static members, 218
statically typed, 17
`static_cast`, 26
`stdafx.h`, 72
strike, 366
`string`, 106
`stringstream`, 107
structurer, 367
subclass, 209
superclass, 209
superclass methods, 216
`switch`, 63

templates, 295
ternary operator, 66
`TEST`, 89
test infected, 91
test-driven development, 91
testing, 85
`this`, 206
threads, 337
`throw`, 61
`true`, 20
`try`, 63
`typedef`, 304

UML, 167, 213, 230
unit test, 85, 245
`unordered_map`, 315
`unsigned`, 19
up-and-out option, 368
user-defined types, 123
`using`, 49

`vector`, 97, 311
vectorisation, 286
version control, 258
`virtual`, 162, 210
virtual destructor, 162, 208, 271
`void`, 45

`while`, 55
World Wide Web, 117

You Aren't Going to Need It
     (YAGNI), 299